D1443515

WILD JUSTICE

Wild Justice

THE MORAL LIVES OF ANIMALS

Marc Bekoff and Jessica Pierce

THE UNIVERSITY OF CHICAGO PRESS : CHICAGO AND LONDON

MARC BEKOFF is professor emeritus of ecology and
evolutionary biology at the University of Colorado,
Boulder. His websites are http://literati.net/Bekoff and
http://www.ethologicalethics.org. **JESSICA PIERCE** is
a philosopher and writer who lives in Colorado.
Her homepage is http://www.jessicapierce.net.

The University of Chicago Press, Chicago 60637
The University of Chicago Press, Ltd., London
© 2009 by The University of Chicago
All rights reserved. Published 2009
Printed in the United States of America

18 17 16 15 14 13 12 11 10 09   1 2 3 4 5

ISBN-13: 978–0-226–04161–2 (cloth)
ISBN-10: 0–226–04161–1 (cloth)

Library of Congress Cataloging-in-Publication Data

Bekoff, Marc.
 Wild justice : the moral lives of animals / Marc Bekoff
and Jessica Pierce.
    p. cm.
 Includes bibliographical references and index.
 ISBN-13: 978-0-226-04161-2 (cloth : alk. paper)
 ISBN-10: 0-226-04161-1 (cloth : alk. paper) 1. Social
behavior in animals. 2. Animal behavior. 3. Motivation
in animals. 4. Animal psychology. I. Pierce, Jessica,
1965– II. Title.
 QL775.B439 2009
 591.5—dc22
                                    2008040173

♾ The paper used in this publication meets the minimum
requirements of the American National Standard
for Information Sciences—Permanence of Paper for
Printed Library Materials, ANSI Z39.48–1992.

MARC DEDICATES THIS BOOK TO HIS PARENTS, WHO TAUGHT HIM THE VALUES OF COMPASSION AND JUSTICE EARLY ON. CLOSE ENCOUNTERS WITH NUMEROUS ANIMALS ALSO TAUGHT HIM THESE VALUABLE LESSONS.

JESSICA DEDICATES THIS BOOK TO THE ANIMALS SHE HAS KNOWN AND LOVED.

CONTENTS

PREFACE INTO THE WILD

It is quite possible that there are . . . a number of intelligent men and women who are not yet aware of the fact that wild animals have moral codes, and that on average they live up to them better than men do theirs.

WILLIAM HORNADAY, *The Minds and Manners of Wild Animals*

A teenage female elephant nursing an injured leg is knocked over by a rambunctious, hormone-laden teenage male. An older female sees this happen, chases the male away, and goes back to the younger female and touches her sore leg with her trunk. Eleven elephants rescue a group of captive antelope in KwaZula-Natal; the matriarch undoes all of the latches on the gates of the enclosure with her trunk and lets the gate swing open so the antelope can escape. A rat in a cage refuses to push a lever for food when it sees that another rat receives an electric shock as a result. A male diana monkey who has learned to insert a token into a slot to obtain food helps a female who can't get the hang of the trick, inserting the token for her and allowing her to eat the food reward. A female fruit-eating bat helps an unrelated female give birth by showing her how to hang in the proper way. A cat named Libby leads her elderly, deaf, and blind dog friend, Cashew, away from obstacles and to food. In a group of chimpanzees at the Arnhem Zoo in the Netherlands individuals punish other chimpanzees who are late for dinner because no one eats until everyone's present. A large male dog wants to play with a younger and more submissive male. The big male invites his younger partner to play and restrains himself, biting his younger companion gently and allowing him to bite gently in return. Do these examples show that animals display moral behavior, that they can be compassionate, empathic, altruistic, and fair? Do animals have a kind of moral intelligence?

We're in an "animal moment." Cornell University historian Dominick LaCapra has claimed that the twenty-first century will be the century of the animal. Research into animal intelligence and animal emotions has come to occupy the agenda in disciplines ranging from evolutionary biology and cognitive ethology to psychology, anthropology, philosophy, history, and religious studies. There is tremendous interest in the emotional and cognitive lives of animals, and there are daily revelations that surprise and even confound some of our assumptions about what animals are like. For example, fish are able to infer their own relative social status by observing dominance interactions among other fish. Fish also have been observed to display unique personalities. We know too that birds plan future meals and that their ability to make and use tools often surpasses that of chimpanzees. Rodents can use a rake-like tool to retrieve food that is out of reach. Dogs classify and categorize photographs the same way humans do; chimpanzees know what other chimpanzees can see, and show better memory in computer games than do humans; animals from magpies to otters to elephants grieve for their young; and mice feel empathy. For anyone who follows scientific literature or popular media on animal behavior it's obvious that we're learning a phenomenal amount.

New information that's accumulating daily is blasting away perceived boundaries between human and animals and is forcing a revision of outdated and narrowminded stereotypes about what animals can and cannot think, do, and feel. We've been too stingy, too focused on ourselves, but now scientific research is forcing us to broaden our horizons concerning the cognitive and emotional capacities of other animals. One assumption in particular is being challenged by this new research, namely the assumption that humans alone are moral beings.

In Wild Justice we argue that animals have a broad repertoire of moral behavior and that their lives together are shaped by these behavior patterns. Ought and should regarding what's right and what's wrong play an important role in their social interactions, just as they do in ours. Even if you feel somewhat skeptical, we ask that you have an open mind and invite you to view animals differently. Indeed, we hope that even the most skeptical readers will come to change their views about the idea of moral behavior in animals.

The term *wild justice* is meant as provocative shorthand. Animals not only have a sense of justice, but also a sense of empathy, forgiveness, trust, reciprocity, and much more as well. In this book we present a unified picture of research concerning moral behavior in animals. We show that animals have rich inner worlds—they have a nuanced repertoire of emotions, a high degree of intelligence (they're really smart and adaptable), and demonstrate behavioral flexibility as they negotiate complex and changing social relationships. They're also incredibly adept social actors: they form intricate networks of relationships and live by rules of conduct that maintain social balance, or what we call social homeostasis.

We also consider the evolution of moral behavior. A cover story in *Time* magazine in December 2007 asked "What Makes Us Moral?" and reviewed the current state of research on the evolution of human morality. In this context the essay gave brief mention to the possibility of moral behavior in animals. If we think that morality has evolved in humans, we're led willy-nilly to ask about its presence in other animals. For a long while there's been agreement that humans and other animals share common anatomical structures and physiological mechanisms. In particular, humans and other mammals have remarkably similar nervous systems.

For readers familiar with evolutionary biology, what we're saying is that arguments for evolutionary continuity—the idea that the differences between species are differences in degree rather than differences in kind—are being supported for a wide variety of cognitive and emotional capacities in diverse species. We believe that there isn't a moral gap between humans and other animals, and that saying things like "the behavior patterns that wolves or chimpanzees display are merely building blocks for human morality" doesn't really get us anywhere. At some point differences in degree aren't meaningful differences at all and each species is capable of "the real thing." Good biology leads to this conclusion. Morality is an evolved trait and "they" (other animals) have it just like we have it.

We also on occasion reference the notion of group selection because our discussion of moral behavior has implications for ongoing debates about individual versus group selection. As we were completing this book a number of articles appeared with catchy titles such as "Survival of the Nicest" and "Survival of the Selfless" in which it was argued that individuals might indeed work "for the good of the group in which they live."

In *Wild Justice*, along with reviewing new research on animals, we offer some larger challenges to how social animals are understood and studied. We challenge the domination—the hegemony, you might say—of the competition paradigm that has monopolized discussions of the evolution of social behavior. The predominance of this paradigm in ethology and evolutionary biology is both misleading and wrong, and momentum is building toward a paradigm shift in which "nature red in tooth and claw" sits in balance with wild justice. The innumerable situations in which we see individual animals working together aren't merely veneers of cooperation, fairness, and trust, but the real thing. Cooperation, fairness, and justice have to be factored into the evolutionary equation in order to understand the evolution of social behavior in diverse species. To this end, we spend a good deal of time discussing social play behavior, an activity that has been overlooked by just about all scholars interested in the evolution of morality. Patterns of behavior observed during play strongly suggest that morality has evolved in animals other than humans.

To support our arguments we consider numerous species in addition to the great apes, especially social carnivores such as wolves. Indeed, even among the great apes there's a good deal of behavioral variation when comparing, for example, chimpanzees and pygmy chimpanzees (bonobos), and this lack of a consistent primate pattern causes trouble for comparative research. We advocate a species-relative view of morality, recognizing that norms of behavior will vary across species. Even within species there might be variations in how norms of behavior are understood and expressed. For example, what counts as "right" in one wolf pack might not be exactly the same as in another wolf pack because of the idiosyncrasies of individual personalities and the social networks that are established among pack members. There isn't one "wolf nature" but rather "wolf natures," just as the renowned biologist Paul Ehrlich argued that there isn't one human nature but rather human natures.

Finally, we argue that the evolution of moral behavior is tied to the evolution of sociality, and that social complexity will be a distinctive marker for moral complexity. We provide examples of nuanced morality when discussing species in which individuals live either predominantly alone or in longlasting social groups in which there are enduring bonds. For example, we'd expect to see more nuanced or fine-tuned morality in packs of gregarious wolves than in less social coyotes and red foxes.

A quick note on terminology. Humans should be proud of their citizenship in the animal kingdom. Yet because of the conventions of the English language, we're apt to forget that humans are animals too. Nonetheless, we use the word *animals* to refer to nonhuman beings because always writing "nonhuman animals" gets tiresome.

Readers may wonder why we're collaborating—Marc Bekoff, a cognitive ethologist, and Jessica Pierce, a philosopher. We first met over roasted artichokes and good merlot at a dinner party thrown by Lynne Sullivan, a mutual friend. We began discussing various aspects of animal cognition and the evolution of moral behavior and it became immediately clear that we had a shared interest, and that collaborating would bring together different fields of expertise and different points of view. As we make clear here, any investigation of the evolution of morality demands discussion and debate across disciplines, and this is precisely what we do. And as we were working on Wild Justice it became clear that people in different disciplines use the same words differently, thus, our collaboration forced us to clarify the jargon that's used to refer to various aspects of social behavior.

We're very excited about our interdisciplinary project and invite others to join us in further developing the study of animal morality, a field that is in its infancy. A mature understanding of the moral lives of animals will require patience and hard work by researchers who are willing to cross disciplinary boundaries and by nonresearchers who share their stories about our moral kin.

The information contained in Wild Justice has profound implications for our moral relationship with, and responsibilities toward, other animals. We will not explore these implications, but we feel that it is important to note that what animals think and feel has to be factored into how we treat them.

Wild Justice travels over hills, into valleys, and around turns. In the first chapter we give an overview of the research on moral behavior in animals. We canvass the social behavior of various species and tell you which we think are the moral animals. We define morality and then sharpen our definition to offer a "species-relative" account of moral behavior.

In chapter 2 we discuss foundations for wild justice, including how scientists make sense of what animals do. We consider the disciplines that have made the most significant contributions to understanding animal morality: cognitive ethology (the study of animal minds), social neuroscience, moral psychology, and philosophy. Researchers in all of these areas have helped to unravel some of the mysteries concerning the cognitive and emotional capacities of animals and how these in turn fold into a discussion of moral behavior. We discuss the use of analogy in science and the value of careful anthropomorphism. We also consider individual and group selection, possible links between intelligence and sociality, and the notion of moral intelligence.

The heart of wild justice is the suite of moral behaviors that fall into three rough "clusters" (groups of related behaviors that share some family resemblances) that we've used as a fulcrum to organize our material: the *cooperation* cluster (including altruism, reciprocity, honesty, and trust), the *empathy* cluster (including sympathy, compassion, grief, and consolation), and the *justice* cluster (including sharing, equity, fair play, and forgiveness). We devote a chapter to each cluster and spell out the evidence for each. At the end of chapter 5 we draw connections among the three clusters to offer a unified picture of the repertoire of moral behavior so as to help readers navigate their way to the conclusion that animals can be moral beings.

In the final chapter, the discussion broadens into philosophy to consider the wider implications of wild justice. Much of this conversation centers on coming to a better understanding of what morality is and what happens when we define it so as to include animals. We also explore the implications of wild justice for sticky philosophical problems such as agency, conscience, relativism, and determinism.

Let's now begin our journey into the world of wild justice. The time has come to put wheels on the discussion of moral behavior in animals so that we can see where we're at and where we need to head in the future. We are not the only moral beings.

FIGURE 1. African elephants walking in a row in Amboseli National Park, Kenya. Elephants are highly social and emotional animals who live in large family groups led by an older, experienced female called the matriarch. Courtesy of Thomas D. Mangelsen/Images of Nature.

1 MORALITY IN ANIMAL SOCIETIES
AN EMBARRASSMENT OF RICHES

Let's get right to the point. In *Wild Justice*, we argue that animals feel empathy for each other, treat one another fairly, cooperate towards common goals, and help each other out of trouble. We argue, in short, that animals have morality.

Both popular and scientific media constantly remind us of the surprising and amazing things animals can do, know, and feel. However, when we pay careful attention to the ways in which animals negotiate their social environments, we often come to realize that what we call surprises aren't really that surprising after all. Take, for example, the story of a female western lowland gorilla named Binti Jua, Swahili for "daughter of sunshine," who lived in the Brookfield Zoo in Illinois. One summer day in 1996, a three-year-old boy climbed the wall of the gorilla enclosure at Brookfield and fell twenty feet onto the concrete floor below. As spectators gaped and the boy's mother screamed in terror, Binti Jua approached the unconscious boy. She reached down and gently lifted him, cradling him in her arms while her own infant, Koola, clung to her back. Growling warnings at the other gorillas who tried to get close, Binti Jua carried the boy safely to an access gate and the waiting zoo staff.

This story made headlines worldwide and Binti Jua was widely hailed as an animal hero. She was even awarded a medal from the American Legion. Behind the splashy news, the gorilla's story was adding fuel to an already smoldering debate about what goes on inside the mind and heart of an animal like Binti Jua. Was Binti Jua's behavior really a deliberate act of kindness or did it simply reflect her training by zoo staff?

Even in the mid-1990s there was considerable skepticism among scientists that an animal, even an intelligent animal like a gorilla, could have the cognitive and emotional resources to respond to a novel situation with what appeared to be intelligence and compassion. These skeptics argued that the most likely explanation for Binti Jua's "heroism" was her particular experience as a captive animal. Because Binti Jua had been hand raised by zoo staff, she had not learned, as she would have in the wild, the skills of gorilla mothering. She had to be taught by humans, using a stuffed toy as a pretend baby, to care for her own daughter. She had even been trained to bring her "baby" to zoo staff. She was probably simply replaying this training exercise, having mistaken the young boy for another stuffed toy.

A few scientists disagreed with their skeptical colleagues and argued that at least some animals, particularly primates, probably do have the capacity for empathy, altruism, and compassion, and could be intelligent enough to assess the situation and understand that the boy needed help. They pointed to a small but growing body of research hinting that animals have cognitive and emotional lives rich beyond our understanding.

We'll never know why Binti Jua did what she did. But now, years later, the amazing amount of information that we have about animal intelligence and animal emotions brings us much closer to answering the larger question raised by her behavior: can animals really act with compassion, altruism, and empathy? The skeptics' numbers are dwindling. More and more scientists who study animal behavior are becoming convinced that the answer is an unequivocal "Yes, animals really can act with compassion, altruism, and empathy." Not only did Binti Jua rescue the young boy, but she also liberated some of our colleagues from the grip of timeworn and outdated views of animals and opened the door for much-needed discussion about the cognitive and emotional lives of other animals.

WILD JUSTICE: WHAT ARE WE REALLY TALKING ABOUT?

Even a decade ago, at the time that Binti Jua rescued the injured boy, the idea of animal morality would have been met with raised eyebrows and a "surely you must be joking!" dismissal. However, recent research is demonstrating that animals not only act altruistically, but also have the capacity for empathy, forgiveness, trust, reciprocity, and much more. In humans, these behaviors form the core of what we call morality. There's good reason to call these behaviors moral in animals, too. Morality is a broadly adaptive strategy for social living that has evolved in many animal societies other than our own.

Our argument relies upon well-established and mostly uncontroversial research. We simply suggest that the many parts, taken together, represent an interesting and provocative pattern. Our most controversial move, of course, is to use the label "morality" to describe what we see going on in animal societies. This jump is controversial not for scientific reasons so much as philosophical ones, and we will keep these philosophical concerns in the foreground of our discussion.

Let us take you through the evidence. We invite you to enter into the lives of social animals. We show that these animals have rich inner worlds—they have a complex and nuanced repertoire of emotions as well as a high degree of intelligence and behavioral flexibility. They're also incredibly adept social actors. They form and maintain complex networks of relationships, and live by rules of conduct that maintain a delicate balance, a finely tuned social homeostasis.

LOOKING FOR THE BAD, LOOKING FOR THE GOOD: THE MORE WE LOOK THE MORE WE SEE

Here's a common distillation of Charles Darwin's theory of evolution. Natural selection, to borrow a popular metaphor from biology, is an evolutionary arms race. Life is a war of all against all, a ruthless and bloody battle, usually over sex and food. Mothers eat their young and siblings fight to the death against siblings (a phenomenon called siblicide). When we look at nature through this narrow lens we see animals eking out a living against the glacial forces of evolutionary conflict. This scenario makes for great television programming, but it reflects only a small part of nature's ineluctable push. For alongside conflict and competition there is a tremendous show of cooperative, helpful, and caring behavior as well.

To offer a particularly striking example, after carefully analyzing the social interactions of various primate species, primatologists Robert Sussman and Paul Garber and geneticist James Cheverud came to the conclusion that the vast majority of social interactions are affiliative rather than agonistic or divisive. Grooming and bouts of play predominate the social scene, with only an occasional fight or threat of aggression. In prosimians, the most ancestral of existing primates, an average of 93.2 percent of social interactions are affiliative. Among New World monkeys who live in the tropical forests of southern Mexico and Central and South America, 86.1 percent of interactions are affiliative, and likewise for Old World monkeys who live in South and East Asia, the Middle East, Africa, and Gibraltar, among whom 84.8 percent of interactions are affiliative. Unpublished data for gorillas show that 95.7 percent of their social interactions are affiliative. After about twenty-five years of research on chimpanzees, Jane Goodall noted in her book *The Chimpanzees of Gombe*, "it is easy to get the impression that chimpanzees are more aggressive than they really are. In actuality, peaceful interactions are far more frequent than aggressive ones; mild threatening gestures are more common than vigorous ones; threats per se occur much more often than fights; and serious, wounding fights are very rare compared to brief, relatively mild ones." These don't appear to be animals whose social lives are defined only by conflict.

The social lives of numerous animals are strongly shaped by affiliative and cooperative behavior. Consider wolves. For a long time researchers thought that pack size was regulated by available food resources. Wolves typically feed on prey such as elk and moose, both of which are bigger than an individual wolf. Successfully hunting such large ungulates usually takes more than one wolf, so it makes sense to postulate that wolf packs evolved because of the size of wolves' prey. However, long-term research by David Mech shows that pack size in wolves is regulated by *social* and not food-related factors. Mech discovered that the number of wolves who can live together in a coordinated pack is governed by the number of wolves with whom individuals can closely bond (the "social attraction factor") balanced against the number of individuals from whom an individual can tolerate competition (the "social competition factor"). Packs and their codes of conduct break down when there are too many wolves.

As we begin to look at the "good" side of animal behavior, at what animals do when they're not fighting each other or committing siblicide, we begin to take in just how rich the social lives of many animals are. Indeed, the lives of animals are shaped at a most basic level by "good"—or what biologists call *prosocial*—interactions and relationships. Even more, it seems that at least some prosocial behavior is not a mere byproduct of conflict, but may be an evolutionary force in its own right. Within biology, early theories of kin selection and reciprocal altruism have now blossomed into a much wider inquiry into the many faces and meanings of prosocial behavior. And, it seems, the more we look, the more we see. There's now an enormous body of research on prosocial behavior, and new research is being published all the time on cooperation, altruism, empathy, reciprocity, succorance, fairness, forgiveness, trust, and kindness in animals ranging from rats to apes.

Even more striking, within this huge repertoire of prosocial behaviors, particular patterns of behavior seem to constitute a kind of animal morality. Mammals living in tight social groups appear to live according to codes of conduct, including both prohibitions against certain kinds of behavior and expectations for other kinds of behavior. They live by a set of rules that fosters a relatively harmonious and peaceful coexistence. They're naturally cooperative, will offer aid to their fellows, sometimes in return for like aid, sometimes with no expectation of immediate reward. They build relationships of trust. What's more, they appear to feel for other members of their communities, especially relatives, but also neighbors and sometimes even strangers—often showing signs of what looks very much like compassion and empathy.

It is these "moral" behaviors in particular that are our focus in *Wild Justice*. Here is just a sampling of some of the surprising things research has revealed about animal behavior and more specifically about animal morality in recent years.

Some animals seem to have a sense of fairness in that they understand and behave according to implicit rules about who deserves what and when. Individuals who breach rules of fairness are often punished either through physical retaliation or social ostracism. For example, research on play behavior in social carnivores suggests that when animals play, they are fair to one another and only rarely breach the agreed-upon rules of engagement—if I ask you to play, I mean it, and I don't intend to dominate you, mate with you, or eat you. Highly aggressive coyote pups, to give just one example, will bend over backwards to maintain the play mood with their fellows, and when they don't do this they're ignored and ostracized.

Fairness also seems to be a part of primate social life. Researchers Sarah Brosnan, Frans de Waal, and Hillary Schiff discovered what they call "inequity aversion" in capuchin monkeys, a highly social and cooperative species in which food sharing is common. These monkeys, especially females, carefully monitor equity and fair treatment among peers. Individuals who are shortchanged during a bartering transaction by being offered a less preferred treat refuse to cooperate with researchers. In a nutshell, the capuchins expect to be treated fairly.

Many animals have a capacity for empathy. They perceive and feel the emotional state of fellow animals, especially those of their own kind, and respond accordingly. Hal Markowitz's research on captive diana monkeys strongly suggests a capacity for empathy, long thought to be unique to humans. In one of his studies, individual diana monkeys were trained to insert a token into a slot to obtain food. The oldest female in the group failed to learn how to do this. Her mate watched her unsuccessful attempts, and on three occasions he approached her, picked up the tokens she had dropped, inserted them into the machine, and then allowed her to have the food. The male apparently evaluated the situation and seemed to understand that she wanted food but could not get it on her own. He could have eaten the food, but he didn't. There was no evidence that the male's behavior was self-serving. Similarly, Felix Warneken and Michael Tomasello at the Max Planck Institute for Evolutionary Anthropology in Leipzig, Germany, discovered that captive chimpanzees would help others get food. When a chimpanzee saw that his neighbor couldn't reach food, he opened the neighbor's cage so the animal could get to it.

Even elephants rumble onto the scene. Joyce Poole, who has studied African elephants for decades, relates the story of a teenage female who was suffering from a withered leg on which she could put no weight.

When a young male from another group began attacking the injured female, a large adult female chased the attacking male, returned to the young female, and touched her crippled leg with her trunk. Poole believes that the adult female was showing empathy. There is even evidence for empathy in rats and mice.

Altruistic and cooperative behaviors are also common in many species of animal. One of the classic studies on altruism comes from Gerry Wilkinson's work on bats. Vampire bats who are successful in foraging for blood that they drink from livestock will share their meal with bats who aren't successful. And they're more likely to share blood with those bats who previously shared blood with them. In a recent piece of surprising research, rats appear to exhibit generalized reciprocity; they help an unknown rat obtain food if they themselves have been helped by a stranger. Generalized reciprocity has long been thought to be uniquely human.

The presence of these behaviors may seem puzzling to scientists or lay readers who still view animals from the old "nature red in tooth and claw" framework. But puzzling or not, moral behaviors can be seen in a wide variety of species in a spectrum of different social contexts. And the more we look, the more we see.

WHAT IS MORALITY AND WHAT MORAL BEHAVIORS DO ANIMALS EXHIBIT?

Before we can discuss the moral behaviors that animals exhibit, we need to provide a working definition of morality. We define morality as a suite of interrelated other-regarding behaviors that cultivate and regulate complex interactions within social groups. These behaviors relate to well-being and harm, and norms of right and wrong attach to many of them. Morality is an essentially social phenomenon, arising in the interactions between and among individual animals, and it exists as a tangle of threads that holds together a complicated and shifting tapestry of social relationships. Morality in this way acts as social glue.

Animals have a broad repertoire of moral behaviors. It's sloppy business trying to squeeze these diverse behaviors into structured categories, but we need some way to organize and present a picture of moral behavior in animals. We envision a suite of moral behavior patterns that falls into three rough categories, around which we have organized our book. We call these rough categories "clusters," a cluster being a group of related behaviors that share some family resemblances, and we identify three specific clusters: the cooperation cluster, the empathy cluster, and the justice cluster. *Wild justice* is shorthand for this whole suite.

The cooperation cluster includes behaviors such as altruism, reciprocity, trust, punishment, and revenge. The empathy cluster includes sympathy, compassion, caring, helping, grieving, and consoling. The justice cluster includes a sense of fair play, sharing, a desire for equity, expectations about what one deserves and how one ought to be treated, indignation, retribution, and spite. We devote separate chapters to exploring each of these clusters in detail (cooperation in chapter 3, empathy in chapter 4, and justice in chapter 5).

Forcing structure in this way raises many questions. Do the behaviors that we cluster together really belong in the same group? For example, is consolation behavior an example of an empathic response, or is it more closely related to cooperation and reciprocity? Are some behaviors more basic than others? For example, is empathy a necessary precursor to fairness? What are the interrelationships between and among behaviors, both evolutionarily and physiologically? Have these behaviors co-evolved? And are we correct in our claim that moral animals will have a behavioral repertoire that spans all three clusters?

WHO ARE THE MORAL ANIMALS?
PENCILING IN A SHIFTING LINE

Many people will immediately want to know who the moral animals are. Can we draw a line that separates species in which morality has evolved from those in which it hasn't? Given the rapidly accumulating data on the social behavior of numerous and diverse species, drawing such a line is surely an exercise in futility, and the best we can offer is that if you choose to draw a line, use a pencil. For the line will certainly shift "downwards" to include species to which we would never have dreamed of attributing such complex behaviors, such as rats and mice.

Taking animal-behavior research as it stands now, there's compelling evidence for moral behavior in primates (particularly the great apes, but also at least some species of monkey), social carnivores (most well studied are wolves, coyotes, and hyenas), cetaceans (dolphins and whales), elephants, and some rodents (rats and mice, at the very least). This isn't a comprehensive catalogue of all animals with moral behavior; it simply represents the animals whose social behavior has been studied well enough to provide ample data to draw conclusions. There are other species, such as many ungulates and cats, for which data are simply lacking. But it would not be surprising to discover that they, too, have evolved moral behaviors.

Research on primates currently provides the most robust account of moral behavior in animals. Given our evolutionary kinship with other primates, it seems reasonable to suppose that these species will have the most behavioral continuity with humans. And indeed, Jessica Flack and Frans de Waal have argued that nonhuman primates are the most likely animals to show precursors of human morality. Yet looking for "precursors" of human morality, though interesting, is not the same as looking for moral behavior in animals. Furthermore, the assumption that primate behavior will be most similar to human behavior may actually prove incorrect. For example, Nobel Prize–winning ethologist Niko Tinbergen and renowned field biologist George Schaller have suggested that we might learn a lot about the evolution of human social behavior by studying social carnivores, species whose social behavior and organization resemble that of early hominids in a number of ways (divisions of labor, food sharing, care of young, and intersexual and intrasexual dominance hierarchies). For these reasons, we're interested in extending the research paradigm on animal morality well beyond primates.

Morality may be exclusive to mammals, and mammals are our focus in this book. At this point, however, it would be premature to pronounce other species lacking in moral behaviors. We simply do not have enough data to make hard and fast claims about the taxonomic distribution among different species of the cognitive skills and emotional capacities necessary for being able to empathize with others, behave fairly, or be moral agents. All must remain quite tentative at this point. It is possible, for example, that some birds, such as the highly intelligent corvids, have a kind of morality. In his book *Mind of the Raven*, biologist and raven expert Bernd Heinrich observed that ravens remember an individual who consistently raids their caches if they catch him in the act. Sometimes a raven will join in an attack on an intruder, even if he did not see the cache being raided. Is this moral? Heinrich seems to think it is. He says of this behavior, "It was a moral raven seeking the human equivalent of justice, because it defended the group's interest at a potential cost to itself." In two subsequent experiments, Heinrich confirmed that group interests could drive what an individual raven decides to do.

There is abundant evidence for the range of behaviors we're exploring in this book, so much so that the basic claim that these behavioral clusters are present to some degree in some animals isn't really controversial at all. But why take the further step and call these behavioral clusters *moral*, a label bound to raise hackles, rather than sticking to the seemingly more objective term *prosocial*?

CHALLENGING AND REVISING STEREOTYPES ABOUT ANIMALS: BAD HABITS ARE HARD TO BREAK

So far, very few scientists and other academics have been willing to use the term *moral* in relation to animal behavior without protective quotation marks (which signal a kind of "wink, wink: we don't really mean 'moral' as in human morality") or without some other modifying trick, as in the term *proto-morality* (read: "they may have some of the seeds of moral behavior, but obviously not morality per se"). Indeed, there is strong resistance to the use of the term "moral" in relation to the behavior of nonhuman animals, both from scientists and philosophers.

The belief that humans have morality and animals don't is such a longstanding assumption it could well be called a habit of mind, and bad habits, as we all know, are damned hard to break. A lot of people have caved in to this assumption because it is easier to deny morality to animals than to deal with the complex reverberations and implications of the possibility that animals have moral behavior. The historical momentum, framed in the timeworn dualism of us versus them, and the Cartesian view of animals as nothing more than mechanistic entities, is reason enough to dismissively cling to the status quo and get on with the day's work. Denial of who animals are conveniently allows for retaining false stereotypes about the cognitive and emotional capacities of animals. Clearly a major paradigm shift is needed, because the lazy acceptance of habits of mind has a strong influence on how science and philosophy are done and how animals are understood and treated.

The irony, of course, is that the field of animal behavior is already bursting with terminology that has moral color: altruism, selfishness, trust, forgiveness, reciprocity, and spite. All of these terms and more are used by scientists to describe the behavior of animals. Certain words like *altruism*, *selfishness*, and *spite* have been ascribed specific and carefully circumscribed meanings within the field of animal behavior—meanings that diverge from, and even sometimes contradict common usage. Other moral terms such as *forgiveness*, *fairness*, *retribution*, *reciprocity*, and *empathy* have joined the animal behavior lexicon, and retain, for now, their connection to the morality we know and live. Lay readers and even scientists are bound to be confused by this apparent lack of consistency. We plan to clear up some of this mess.

We could have coined a new word or phrase to describe our particular suite of prosocial behaviors in animals. The phrase "animal morality" will certainly strike some people as odd, and perhaps even as an oxymoron. And in some respects, *morality* is not the most solicitous term. Morality is notoriously hard to define and there is disagreement about how best to understand what morality is. On the other hand, morality is a very useful term, because "animal morality" challenges some stereotypes about animals and, as we'll see, about humans. It also emphasizes evolutionary continuity between humans and other animals, not only in anatomical structure, but also in behavior. And this emphasis, in our view, is important. Finally, *morality* is also a useful term because the root meaning—*more*, or custom—captures an essential element of animal morality.

We need to be quite explicit that the meaning of morality is itself under consideration, and we're suggesting a shift in meaning. How we define morality will, of course, determine whether and to what extent animals have it. And yes, we're defining morality in such a way as to lend credence to our argument for evolutionary continuity between humans and animals. But this is not sleight of hand: our definition of morality is well supported both scientifically and philosophically and also by "unscientific" common sense. We want to detach the word *morality* from some of its moorings, allowing us to rethink what it is in light of a huge pile of research from various fields that speaks to the phenomenon. We ask that you let us play freely with the term and, in the end, you can decide if you think "animal morality" makes sense.

MORALITY AND PROSOCIALITY: CLARIFYING CATEGORIES

The animal behavior literature tends, as we noted, to avoid the word *moral*, using instead the more neutral and technical-sounding *prosocial behavior* (actions that benefit another individual) or more specific terms such as *altruism, empathy,* or *cooperation*. This term *prosocial* is of central importance to us as we explore the distribution of moral behavior among animals. *Prosocial* is used in the literature on animal behavior to describe many of the behaviors that we want to call *moral*. Unfortunately, *prosocial* does not seem to have a clear, unambiguous definition and is used in a variety of ways, sometimes as a synonym for altruism, sometimes for cooperation, sometimes for succorance, sometimes for empathy, and sometimes for a rather vague conglomeration of these behaviors.

Moral and *prosocial* are intimately linked and overlapping concepts, but they are not synonyms. As far as we know, there has been no careful delineation of the prosocial in relation to the moral, either for humans or for animals. If *moral* is a term that becomes part of the lexicon of ethology, as we hope it does, then careful work must be done to distinguish the two. We offer an initial proposal here, and invite dialogue.

Morality and prosociality represent distinct categories, though with considerable overlap. In evolutionary terms, prosocial behavior is at the root of morality, and is much more broadly distributed than morality. Many prosocial behaviors would fall outside the narrower "moral" category. For example, parental care and communal nursing are not, in themselves, moral behaviors. Neither is altruism, as understood in the scientific literature, behavior in which the actor provides another individual with a benefit, but in doing so incurs some cost, where cost and benefit are understood in terms of future reproductive success. The self-sacrificial behavior of ants, bees, and wasps does not constitute morality, nor does sentinel behavior, where animals take turns watching for predators.

Thus many species in which prosocial behavior is displayed do not have moral behavior. Ants and bees behave prosocially, but not morally. Why would we say that wolves have morality, while ants don't, even though both species engage in cooperative and altruistic behavior? We propose certain threshold requirements for given species to have morality: a level of complexity in social organization, including established norms of behavior to which attach strong emotional and cognitive cues about right and wrong; a certain level of neural complexity that serves as a foundation for moral emotions and for decision making based on perceptions about the past and the future; relatively advanced cognitive capacities (a good memory, for example); and a high level of behavioral flexibility. We explore these threshold requirements for morality in more depth in later chapters.

Most moral behavior could be also classed as prosocial. But some behavior might be considered moral, even though it is not technically prosocial. For example, behavior aimed at avoiding harm to another might fall into the category of morality, but not prosociality, since we've defined prosocial behavior as that which promotes the welfare of others (whether intentionally or not). Of course, not all behavior that avoids causing harm should be classed as moral, either. But where the avoidance of harm to another is other-regarding, where it is motivated by a desire to get along with others in one's society, it should be considered moral behavior.

FINESSING MORALITY: PROHIBITIONS AND PROSOCIALITY

Social animals live according to well-developed systems of prohibitions *against* certain kinds of behavior and proscriptions *for* certain kinds of behavior. These prohibitive and proscriptive norms govern the behavior of individuals within a group and relate to harm, welfare, and fairness. These behaviors, in philosophical lingo, are *other-regarding*, as opposed to *self-regarding*. A self-regarding action affects no one other than the agent (the individual) performing the action. An action or behavior becomes other-regarding when it produces some benefit to another, causes some harm, or violates some social rule or obligation—basically, when it affects the welfare of another individual or the social group. There might be prohibitions against certain kinds of harm, both physical (biting, killing, violent aggression) and psychological (bullying, taunting, intimidating), under certain circumstances. There may also be expectations about helping, reciprocating, and sharing. Within an animal society, for example, there might be certain norms of reciprocity: help those who have helped you (you owe them) and help those who need help (regardless of payoff). And there might be norms about fairness: those with highest status eat first and best, and those who invite play should follow the rules of play. Norms might govern and maintain dominance hierarchies, regulate the acquisition and distribution of food, regulate grooming behaviors, regulate sentinel behaviors, or govern play behavior. (A norm is an expected standard of behavior within a group and is enforced by the group.) Harm and benefit are the basic units of moral currency.

Beneath these proscriptions and prohibitions lies the raw material of a sympathetic species. Social animals have well-developed instincts, such as a range of empathic behaviors, which help to create and maintain a culture of fellow feeling. Recent research indicates that prosocial behaviors such as empathy and reciprocity have both cognitive and emotional elements, though how these relate is still an open question. Research on animal behavior, combined with research into human psychology and neuroscience, can help elucidate some of the underlying mechanisms at play.

MORALITY AND MANNERS

When you see a child do something especially rude you might roll your eyes and say, "He must have been raised by wolves." By human standards, a child behaving like a wolf is quite ill mannered. But in wolf society, it's just fine to stick your face in your food (or elsewhere), growl, and gulp down as much as you can in ten seconds flat. Wolf manners are actually quite good, if you're a wolf.

Like morality, manners regulate social behavior. Researchers in the field of human moral psychology have given a great deal of attention to the distinction between moral violations and conventional violations. Conventional violations, it is said, are wrong by standards of social acceptability. Moral transgressions are more serious, and their wrongness relates to harming others. Driving on the correct side of the road or eating salad with the shorter fork have little to do with fairness, reciprocity, or the welfare of others.

Animals could certainly be said to have manners as well as morality. There are species-specific rules about who eats first, and about proper methods of grooming or making introductions. It's also likely that in animal societies "manners" such as grooming and eating queues have strong moral importance—these are part of the social conventions that help maintain group cohesion and cooperation. We speculate that the distinction between manners and morals (or social conventions and moral conventions) may be less pronounced in animal societies than in human societies.

In philosophical discussions, human morality is often compared not only to etiquette but also to law and religion. Law usually has considerable overlap with morality, but is governed by explicit rules and punishments, whereas morality is an informal system of behavioral control. Religion, of course, invokes supernatural explanations for why certain behaviors are prohibited or required. It seems likely that morality (with manners as a subset) is really the only category that applies to nonhuman animals.

NASTY NICK: MORALITY AND IMMORALITY ARE TWO SIDES OF THE SAME COIN

Turn of the century zoologist William Hornaday wrote in *The Minds and Manners of Wild Animals: A Book of Personal Observations*, "The animal world has its full share of heroes. Also, it has its complement of pugilists and bullies, its cowards and its assassins." And he's right. On occasion, animals aren't nice to one another. Take, for example, the olive baboon Nick, both pugilist and bully. Nick was an adolescent when he joined what was known as the Forest Troop in the southeast corner of Masai

Mara National Reserve in Kenya. You could almost see contempt on his face, according to renowned Stanford University professor Robert Sapolsky, who wrote about Nick in *A Primate's Memoir*. Sapolsky noted that Nick dominated his age group, and that "he was confident, unflinching, and played dirty." Sapolsky is known for speaking eloquently and plainly about animal behavior and he's equally blunt about Nick: "The guy simply wasn't nice . . . He harassed the females, swatted at kids, bullied ancient Gums and Limp." In one instance, Nick trounced a baboon named Rueben in a fight. Rueben "stuck his ass up in the air," a signal of submission and vulnerability that should have ended the dispute. Nick, however, used this as an opportunity to slash Rueben's butt with his sharp canines, in clear violation of baboon social norms.

The story of nasty Nick points to an important question: can animals be immoral? We say yes. The formula is actually quite simple. In those animal species where we find moral behavior we also expect to find immoral behavior. Moral and immoral need each other like peanut butter and jelly; you won't find one without the other.

Just as we don't want to understand any and all behavior that benefits another as moral (we don't want to say that helper ants are moral), we wouldn't want to define any and all behavior that harms another as immoral. It's ridiculous to suggest that the lion hunting down and killing the deer is immoral, however ruthless his behavior seems in Mutual of Omaha's *Wild Kingdom*. Nor is an egret pecking its sibling to death an instance of bad upbringing. Nor, to offer one final example, is "dishonest signaling," such as when a male frog "lies" about his prowess by maintaining a close proximity to the loudest croaker, thus hoping that a female will make a mistake and believe him to be the source of that alluring music.

Behavior becomes immoral when it goes against socially established expectations. During predation, there's no prior agreement by the wolf not to eat the elk; there are no social expectations, since wolves and elk don't live in the same society. So, there's no violation of a social norm. On the other hand, if two wolf pups are playing and one tries to dominate the other, a norm has been violated.

In animals with the capacity for moral behavior it seems reasonable to conclude that the cognitive and emotional skills that underlie morality might be put to use in antisocial as well as prosocial ways. For example, Frans de Waal points out that empathy rests upon the capacity to understand others, in particular their suffering, and this capacity makes cruelty possible. Empathy and cruelty both rely on the ability to imagine how one's own behavior affects others. We know how to cause pain and distress. The same logic applies to other behaviors. Trust and honesty form the glue in cooperative social groups. Yet a reliance on trust is what makes deception and dishonesty possible. In cooperative groups, deception is always a successful strategy, but it is less successful, on the whole, than cooperation.

Let's consider the question of cruelty in animals in a bit more detail, because discussions of the rare instances of animals being cruel to each other are often overinflated and generalized, and presented as confirmation of the "nature red in tooth and claw" paradigm. Available data are actually quite slim because of small sample sizes and a good deal of variability among different communities of animals. For example, in their 2006 review of comparative rates of violence in chimpanzees and humans Harvard anthropologist and chimpanzee expert Richard Wrangham and his colleagues Michael Wilson and Martine Muller note, "the relatively small sample size and great variation among sites renders imprecise any estimate of violence-related mortality rates for chimpanzees as a species."

Animals surely have the capacity to be cruel, but our reading of available data indicates that they rarely express it. Because outright cruelty is rare it captures our attention when it occurs. However, it's misleading to claim that cruelty trumps affiliative or neutral social interactions in the long run. For example, when a dog invites another dog to play and then beats it up it's an attention-getter, but in fact these sorts of interactions are extremely rare, even among dog's wild relatives. Many people know about Jane Goodall's single observation of a group of male chimpanzees pursuing and then killing all the members of another chimpanzee group over a two-year period. Goodall described this behavior as warlike, and was shocked by the intentional brutality of the chimpanzees. A lot of people use the Gombe war incident, the relatively rare occurrence of infanticide (for example, adult male lions killing baby lions in order to encourage a female to become reproductively active), and the occasional harassment and beating of a low-ranking scapegoat wolf to argue that nonhuman animals have the capacity to be cruel. Others, however, are reluctant to use isolated and rare examples of what appears to be cruel behavior to generalize about cruelty across animal species.

Psychologist Victor Nell has argued that cruelty is an exclusively hominid behavior. "Cruelty (from the Latin crudelem, 'morally rough') is the deliberate infliction of physical or psychological pain on a living creature; its most repugnant and puzzling feature is the frequently evident delight of the perpetrators." He believes that cruelty is a behavioral by-product of predation. Cruelty was adaptive to our ancestors because it led to successful predation, and it was (and still is) reinforced through affectively positive neurobiological mechanisms—in other words, cruelty feels good. Nell believes that seemingly cruel behaviors such as cat-and-mouse play, or orcas "playing" with baby seals before eating them, are most parsimoniously interpreted as extreme forms of aggression. Animals, in his view, are not imagining, much less enjoying, their victim's suffering. Cruelty requires certain cognitive capacities, such as the intention to inflict pain (which, in turn, presupposes a theory of mind), and he does not believe that animals can reflectively imagine the suffering of another.

Nell's paper "Cruelty's Rewards" generated a lively debate among ethologists and others. Some scientists took issue with Nell's claim that only humans can be cruel. They offered various counterexamples, and evidenced a rich literature on cruelty in nonhuman primates and other mammals. The debate about cruelty in animals is clearly relevant to wild justice, particularly since it involves understanding whether and to what extent animals have theory of mind and other advanced cognitive skills. This will be another avenue for fruitful comparative research. However, because of its extreme rarity, we may forever be dependent on stories of animal cruelty rather than on large databases. Ultimately, wild justice does not stand or fall on the question of cruelty. Animals can be moral, with or without the capacity for cruelty.

Groups of social animals have systems in place for dealing with violations of the moral code. These sanctioning mechanisms are a good way to identify and understand what, in any given animal society, is considered immoral behavior. Violations might consist of being overly aggressive or domineering, of refusing to share appropriately, or of being a freeloader, liar, or cheater. In the context of play behavior, for example, violations of the moral code would include accepting an invitation to play and then violating the rules of play by biting too hard or trying to mate, which run counter to expected behavior. Sanctioning behaviors include physical punishment, social ostracism, and future payback (for example, a coyote refusing to play or to share in future encounters). In species such as chimpanzees, in which reciprocity and fairness figure large, there are punishments for those who break the rules. Those chimpanzees who fail to share appropriately are treated with less generosity by others, and they may be ostracized. To understand "fairness" or "reciprocity," animals must have some understanding of their opposites.

We need to be careful not to make "moral" (or "altruistic" or "prosocial") the opposite of "selfish." It isn't. Many moral behaviors are motivated by self-concern, broadly understood. We conform to norms of behavior because otherwise we face social sanctioning, in the form of ostracism, embarrassment, shame, and payback.

MORALITY'S CONTINUUM: A SPECIES-RELATIVE ACCOUNT

We advocate a species-relative view of morality. Each species in which moral behavior has evolved has its unique behavioral repertoire. The same basic behavioral capacities will be present—empathy, altruism, cooperation, and perhaps a sense of fairness—but will manifest as different social norms and different behaviors (e.g., different grooming patterns or unique ways of expressing empathy). Despite some shared evolutionary history, wolf morality is different from human morality and also from elephant morality and chimpanzee morality.

Comparative research on animal morality can be quite valuable, but species differences must also be borne in mind because each species is unique. In particular, comparisons between humans and other mammals should be extremely cautious. We should especially resist the temptation to use humans as the gold standard by which to judge the morality of nonhuman species. In other areas of comparative biology (e.g., auditory and olfactory communication), the human-as-gold-standard has proven deficient because each species has its own distinctive capacities adapted to its own particular environmental and social circumstances. Renowned biologist Edward O. Wilson places humans in a category distinct from other social vertebrates; he does this, presumably, because human sociality is so unique. Humans have achieved a level of social complexity unparalleled in other species. We've also developed the most complex and nuanced morality, and articulate and communicate norms using symbolic language. If we assume that morality in other species will look just like human morality, we're likely to conclude that they don't have morality, having blinded ourselves to this fascinating aspect of their behavior. Rather, we need to proceed with an open mind and view each species on its own terms.

We should also bear in mind that even within a species there can be considerable variation. One society of Animal X may not behave exactly like another society of Animal X, and within each society of Animal X there are unique individuals, each with their own particular personality and life experience. For example, not all populations of chimpanzees use tools, and erroneous conclusions would have been drawn if chimpanzee research didn't involve observations of different populations. Think especially of how wrong we would have been if studies of chimpanzees were only conducted on groups in which tool use didn't occur. We'd still be referring to humans as *Homo faber*, the tool user, to indicate that only humans had evolved the skills necessary to manufacture and use tools. Jane Goodall's observations in the early 1960s of David Greybeard using a blade of grass to get termites out of a hole showed just how misleading this conclusion would have been.

Not only will the set of actions that constitute moral behaviors vary among species, but so will the degree of moral complexity vary from species to species. We propose as an invitation to further discussion that in those animal species with evolved moral behaviors there can be varying degrees of sophistication and complexity. Morality is not an all-or-none phenomenon. Rather, it is nuanced. Animals with a highly developed moral capacity may include chimpanzees, wolves, elephants, and humans. In these species we see a wide-ranging suite of complex social behaviors. Emotions are rich and varied. Facial displays are subtle and carry social meaning. There is evidence in these species of complex cognitive empathy (trying on the perspective of someone else) and not merely emotional contagion (responding automatically to another's emotional state; I'm scared because you're scared).

Rats and mice seem to have a less sophisticated moral repertoire than wolves or chimpanzees. We know from research done in the 1960s that rats will not take food if they know that their actions cause pain to another rat, and recent work with mice likewise shows a capacity for empathy. We also know that rats and mice live in cooperative social groups, are quite intelligent, and experience a range of emotions. Nonetheless, their moral capacities seem to be less complex than those of chimpanzees and humans. Research on mouse empathy, for example, suggests a capacity only for a relatively simple and reflexive form of empathy called emotional contagion. On the other hand, there hasn't been detailed study of rat morality or mouse morality, so we could be surprised. Indeed, a newly published Swiss study showed that rats exhibit what is called "generalized reciprocity"—they generously help an unknown rat obtain food if they themselves have benefited from the kindness of a stranger. Continued research on rat sociality may force us to revise our generally dismissive and disgusted attitude toward these animals.

BIOLOGICAL DETERMINISM AND MORALITY: DO GENES RULE?

Speaking of "lowly" rodents brings us to one final, important point: the role of genes and experience in shaping behavior, the old nature-versus-nurture debate. E. O. Wilson argued, first in his 1975 seminal and controversial book *Sociobiology* and then more fully in his 1978 Pulitzer Prize–winning work *On Human Nature*, that genes determine not only the physical features of an organism but also its behavior. Even moral behavior is genetically wired. Sociobiology soon became a label to describe a new academic discipline, and also, in a sense, a new school of social thought: it described a particular way of understanding how biology relates to social behavior. Although sociobiology did little more than spell out the implications of neo-Darwinian thought in the realm of behavior, many people considered sociobiology dangerous because they saw in it the modern rebirth of social Darwinism. And people feared the resurrection of ideas that were used to justify phrenology, eugenics, and other forms of racial imperialism and genetic determinism.

Some may worry that our ideas are similarly dangerous because we argue, like Wilson did, that morality is at least in part a product of genes. But these fears are misplaced. As we note in this chapter and elsewhere, having the genetic wiring for a particular behavior such as empathy says little about how this behavior will be expressed or about its modifiability or flexibility. Whether or not, and to what extent empathy is expressed depends on a number of factors: what happens during early development, parental influence, social and environmental context, experience, and so forth. It is worth reminding ourselves that the nature/nurture dichotomy is generally considered dead: the consensus among scientists is that behavior is shaped by a complex interplay of many factors.

Some people fear an evolutionary account of morality because they believe that it reduces morality to "mere" biological mechanisms, so that parental love, the loyalty of friends, and the generosity of strangers is reduced to genetic hardwiring. At the same time, immorality—rape, aggression, even war—is reduced to "natural urges" and thus excused or even justified. But this reductionism, though plenty of examples of it can be found in the literature of sociobiology and evolutionary psychology, does not follow, ex post facto, from the scientific evidence. Seeing the biological roots of morality doesn't mean that we then have to excuse malicious or evil behavior—it is still malicious and evil. Likewise, love, loyalty, and generosity are all very real. Does morality have a biological basis? The answer is most certainly yes. This doesn't mean, however, that biology is all there is to say about morality, or that biology somehow has the last word.

Oscar Wilde is said to have remarked, "Morality, like art, means drawing a line someplace." Many of the ideas we're proposing in this book are controversial, and much is still unknown about the moral lives of animals. We're not at all sure if we're right, but we think it's useful to pencil in some lines (e.g., "morality is this but not that"; "these animals but not those probably have morality") and have a good eraser in hand.

These lines are a tool in that they allow for critical, focused discussion about what morality is, who has it, and why we should care.

We invite you to come along on our journey into the moral lives of animals. Before we look at moral behavior in detail, we lay some groundwork in the next chapter for understanding the science behind our arguments.

2 FOUNDATIONS FOR WILD JUSTICE

WHAT ANIMALS DO AND WHAT IT MEANS

We admit that our project is highly ambitious and perhaps controversial, so it's important to lay out clearly where we're coming from. We plan to show that the ice on which we're traveling isn't really all that thin.

Our first project here is to describe the disciplinary foundations for animal morality. The substantial body of research that supports our views on wild justice stems from different fields, particularly cognitive ethology, social neuroscience, moral psychology, and moral philosophy. Although these are different areas of study, there is remarkable overlap in the search to understand moral behavior, both in humans and other social animals. Indeed, the concept of animal morality unifies a number of diverse threads of research into an interesting whole.

Our second project is to provide an overview of our methodological framework. We'll describe how data on animal behavior, especially information that casts light on social relationships and individual variation, is collected, analyzed, and interpreted, and what types of data we need to show that animals behave altruistically, empathically, or fairly. We'll explore several methodological challenges in studying animal social behavior, such as the "problem" of anthropomorphism (attributing human characteristics to animals) and the potential dangers of drawing analogies between human and animal behavior. And we'll also consider various scientific and philosophical debates about animal minds, such as whether animals have "theory of mind" and how the privacy of mental experience limits what we can know about the cognitive or emotional lives of animals.

Finally we'll give a brief overview of the "raw materials" of animal morality: sociality, intelligence, and emotion. Morality arises out of sociality and is intimately tied to both intelligence and emotion. Indeed, morality can be seen as a kind of intelligence all its own, one that weaves together cognitive skills (memory, predictions about the future behavior of others) and emotional skills (the ability to "read" facial expression, body postures, olfactory cues, social dynamics) into a unique kind of social intelligence. Morality, as a behavioral repertoire, draws upon and seems to unify diverse skills and capacities.

COGNITIVE ETHOLOGY: THE STUDY OF ANIMAL MINDS AND WHAT'S IN THEM

Our argument draws on and unifies information gathered from a large number of scientific fields as well as from the humanities, but we rely primarily on research in cognitive ethology. Cognitive ethology is a branch of ethology, which is the study of animals in their natural settings or in situations closely resembling them. Ethologists study various facets of animal behavior, including patterns of communication, aggression, sexual behavior, cognition, learning, emotions, and culture. Incidentally, the term *ethology* is derived from the Greek *ethos*, meaning "custom"—also the root of the word *ethics*. Cognitive ethologists are interested in mental continuity among different species and in comparing thought processes, consciousness, beliefs, and rationality in animals. Some also want to know how and why the intellectual, emotional, and moral skills of animals evolved, all in an attempt to understand the animals themselves, including individual differences, the behavior of social groups, and variations among species.

Cognitive ethologists typically follow the methodological framework set forth by Niko Tinbergen, one of the early pioneers of the field. Tinbergen's contributions to the field of animal behavior were so important that he was awarded the Nobel Prize in 1973, an honor he shared with Konrad Lorenz, author of *On Aggression*, who also wrote on many aspects of animal behavior including imprinting, and Karl von Frisch, who discovered bee language and wrote the illuminating book *The Dance Language and Orientation of Bees*. Tinbergen identified four overlapping areas with which ethological investigations should be concerned, whether a researcher is interested in how herring gulls avoid being eaten by red foxes, how wasps find their homes after hunting excursions, how geese court one another, how dogs play, or how elephants comfort one another. He suggested that researchers should be interested in (1) the *evolution* of a behavior; (2) *adaptation*, or how the performance of a specific action allows an individual to fit into his or her environment and ultimately allows him or her to breed; (3) *causation*, or what causes a particular behavior to occur; and (4) *development* or ontogeny, how a behavior arises and unfolds over the course of an individual's life, giving rise to individual differences.

For example, if we're interested in how and why dogs play, we want to answer the following four questions: (1) Why has play evolved in dogs, and why has it evolved in some animals, like dogs, but not in others? (2) How does play allow a dog to adapt to his or her environment, and how does it influence his or her reproductive fitness? (3) What causes dogs to play? For example, what stimuli elicit play behavior (e.g., the play bow)? (4) How does play behavior develop in young dogs and how does the behavior change as individuals grow older?

Ethologists often also talk about giving *ultimate* and *proximate* explanations for a particular behavior. An ethologist might be interested in an *ultimate* explanation for a behavior, seeking to understand why, for example, play has evolved and how it contributed to the reproductive fitness of an individual wolf. Tinbergen's first two research questions seek ultimate explanations. An ethologist might also, or instead, look for what are called the *proximate* explanations: What immediate goal is an individual pursuing and what internal mechanisms are guiding its behavior? What are the cognitive and affective underpinnings of the behavior? What is the trigger stimulus? For example, a proximate trigger might be a play invitation signal given by one wolf to another. Tinbergen's third and fourth questions relate to proximate explanations for behavior in that they look for what's going on *now*, in the immediate social context, not in the evolutionary past. The two types of explanation are obviously closely interconnected, but it is essential to keep in mind which type of explanation one is after.

STUDYING BEHAVIOR: OBSERVING
AND RECORDING WHAT ANIMALS DO

In the early days of ethology people were unsure how to observe and measure behavior because it just happens and disappears, but Konrad Lorenz stressed that behavior is something that an animal "has" as well as something he or she "does." It can be thought of in the same way in which we think of an anatomical structure or organ on which natural selection can act. With careful study we can describe an action just as we would a heart or a stomach; we can measure the action and learn why animals perform certain behavior patterns in certain situations.

So, the basic research method for answering Tinbergen's questions involves careful observation and description of the behavior patterns performed by the animals being studied. The information provided by observations allows a researcher to exploit the animal's normal behavioral repertoire to answer questions about the evolution, function, causation, and development of the behavior patterns that are exhibited in various situations. Since behavioral abilities have evolved in response to natural selection pressures, cognitive ethologists favor observations and experiments on animals in conditions that are as close as possible to the natural environment where selection occurred. However, the study of captive animals (especially in conditions that closely mimic the natural environment) can add valuable data that simply cannot be gathered in the field, such as the dynamics of social interactions of secretive animals like solitary cats, or of young animals in and around their nest or den.

MORALITY ON THE BRAIN: ADDING SOCIAL
NEUROSCIENCE TO THE PICTURE

Research in social neuroscience, which explores the biological foundation of social behavior, particularly how the brain and nervous system influence social behaviors such as affiliation, empathy, or trust, adds additional color and detail to ethological discoveries about the social, emotional, and moral lives of animals, and continues to demonstrate how strong and widespread the continuities between humans and animals are. Recently, the renowned neurobiologist Donald Pfaff, working at the Rockefeller University, published an entire book devoted to the neuroscience of fair play and altruism. In this book Pfaff argues that the "golden rule" is hardwired into the human brain. Research by Jorge Moll and his colleagues is providing numerous insights into the neural basis of human morality and altruism.

While ethology relies largely on data drawn from the observation of behavior, social neuroscience tends to look for proximate mechanisms or immediate causes of behavior, trying to find, for example, the neural correlates (which areas of the brain are activated) and physiological processes (which hormones are released into the brain) related to empathy or trust or some other social behavior. As an example of social neuroscience research related to animal morality, the work of neurobiologist Jaak Panksepp has been very important in providing insight into the social behavior of rats. Rather than observing rat interactions in the wild, as an ethologist might, Panksepp explores what is happening inside the brains and bodies of rats by carefully orchestrating particular kinds of social interactions in the lab, and then taking thin slices of brain tissue from the rats and detailing patterns of neural activity. Panksepp has made particularly important discoveries about the neurochemical mechanisms underlying emotions. For example, his research has shown that play behavior in rat pups causes the release of opioids into the brain, producing feelings of social comfort and pleasure. He has also discovered that rats experience joy and even that they laugh when tickled.

Two areas of current research in social neuroscience with huge potential to contribute to our understanding of animal morality are mirror neurons and spindle cells. Mirror neurons were first discovered, more or less by accident, in the early 1990s. Researchers studying the areas of the brain involved in hand movement were monitoring the brain activity of macaque monkeys as they picked up pieces of food. They noticed that certain neurons fired when the monkeys watched the researchers pick up food—the same neurons that fired when the monkeys picked up food themselves. The monkey's brains were "mirroring" the researchers' movements.

In November 2007 scientists reported that individual mirror neurons exist in human brains. And, it turns out, humans have neurons with mirror properties widely distributed throughout our brains. These neurons allow us to understand another individual's behavior by imagining ourselves performing the same behavior and then mentally projecting ourselves into the other individual's shoes. Scientists now believe that mirror neurons in humans may play a role in the development of language and, of special relevance to this book, in the ability to understand the emotions of others. Just as the brain mirrors motor movements, it also mirrors emotions. Mirror neurons may thus be key to understanding empathy—our capacity to share the feelings of another. In 2006, mirror neuron researcher Giacomo Rizzolatti was quoted in the *New York Times*: "Mirror neurons allow us to grasp the minds of others not through conceptual reasoning but through direct simulation. By feeling, not by thinking." Researchers believe that mirror neurons might also be used in other modalities such as hearing and smelling. And deficiencies with the mirror neuron system may underlie cognitive disorders such as autism. Neuroscientist V. S. Ramachadran claims that "mirror neurons will do for psychology what DNA did for biology" because they will provide a unifying framework for understanding a whole range of mental abilities. Although this conclusion may be overdrawn, there's no doubt that the discovery of mirror neurons is a landmark achievement that will shape future research into human and animal minds.

Comparative work on mirror neurons is still in its infancy. Mirror neurons have also been observed in birds, where they might play a role in the imitation of sounds. Mirror neurons might also explain observations of empathic mice who react more strongly to painful stimuli after observing other mice in pain, of rats who go hungry rather than watch another rat receive a shock, and of rhesus monkeys who won't accept food if another monkey suffers when they do so.

Another important discovery in neuroscience is the presence of spindle cells (also called von Economo neurons) in whales. Previously, researchers had assumed that only humans and other great apes had these specialized and very large neural cells, which appear to play a role in empathy, intuition about the feelings of others, as well as in rapid gut reactions. In 2006, Patrick Hof and Estel van der Gucht reported the presence of spindle cells in humpback whales, fin whales, killer whales, and sperm whales in the same area of their brains as spindle cells in human brains. Spindle cells in whales are found in the anterior cingulate cortex and frontoinsular cortex, two areas of the brain that are important in reactions that require rapid emotional judgments, such as deciding whether another animal is in pain and the feeling of whether an experience is pleasant or unpleasant. Whales, it turns out, have three times more spindle cells than humans. To sum up the significance of spindle cells in whales, Lori Marino, a cetacean expert at Emory University, notes, "this is consistent with a growing body of evidence for parallels between cetaceans and primates in cognitive abilities, behaviour, and social ecology."

While the data generated by social neuroscience are extremely valuable for learning more about animal minds, these studies are especially troubling because of the pain and suffering that individual experimental animals endure. We mention this because the more we understand about animal cognition and emotions, the more ethically problematic this sort of research becomes.

AND A LITTLE BIT OF PHILOSOPHY

Wild Justice isn't primarily a book of philosophy, but philosophy is certainly important for what we want to do. Indeed, philosophy is always relevant to science: science is shaped in important ways by the worldview of those who engage in it. Our philosophy (broadly understood) shapes the sorts of questions we're liable to ask and the types of answers we'll be open to finding. But our book intersects more pointedly with philosophy than most.

"Do animals have moral behavior?" is a question that is neither pure science nor pure philosophy, and really we have to address both aspects of the question at once. There are, on the one hand, legitimate scientific questions about what exactly is going on in the minds of individual animals and in the complex social interactions among a community of group-living animals, and these have a central bearing on whether we can appropriately describe some animal behavior as moral. These questions center on the capacity of animals to experience rich and complex emotions, on whether animals have self-awareness, remember past events, predict the future, and "understand" complex social interactions in complex ways. These questions also ask us to pay attention to the subtle nuances of interrelationships, of what happens *between* and *among* animals. We argue that the data support our assertion that certain behavioral patterns in animals constitute a system of morality, and that over time scientific resistance to using the term "animal morality" will dissipate.

In making a case for moral behavior in animals, we're also addressing the much bigger question "What is morality?" Let us be very clear about our agenda: we are interested in the behavior of animals, and are not trying to do a comparative analysis of human and animal morality. But in exploring the phenomenon of moral behavior, which is shared in our view by all highly social mammals (including humans), we cannot avoid some discussion of human moral behavior. Indeed, "What is morality?" has been answered thus far only in relation to *Homo sapiens*, and so we cannot avoid some attention to how human morality has been understood.

Research on human moral behavior, carried out over the past several decades in philosophy, converges with the animal data in interesting ways. The answer to the question "What is morality?" has been shifting and evolving. In many ways the research shows that human moral behavior is much more "animal-like" than our common-sense assumptions would suggest. For example, morality has generally been equated with rational judgment and action—we're faced with a moral dilemma, we make a judgment (based on moral principles, ideally) about the best course of action, and then we act. Yet it turns out that reason and judgment don't so seamlessly connect with action. Work in human psychology has shown that context (the specifics of a situation) strongly affects or biases action, so much so that "judgment" is in no sense pure. Philosophers John Doris and Stephen Stich offer several examples from the annals of social science. In one study, subjects who had just found a dime were twenty-two times more likely to help a woman who had dropped some papers than those who had not found a dime. Another study found that subjects were less likely to help an injured man who had dropped some books when a power lawnmower was running nearby than when ambient noise levels were normal.

It seems that at least some moral behavior is "hardwired" into our very physiology. Morality is a product of biological traits that have evolved, and recent research in cognitive neuroscience is discovering the physical correlates of the moral sense. Human capacities such as empathy and justice and trust are physical processes involving the brain, as well as other bodily systems. For example, studies have shown that when levels of the hormone oxytocin increase, a willingness to trust also increases. This is an unconscious, involuntary response. It does not rely on higher cognitive processing. The empathy response can be similarly involuntary (though it can also be shaped by cognition). These processes arise in response to the environment, especially the social environment. Our brains are constantly plugged in to the social network.

We believe that the most appropriate definition of morality is an expansive one that includes under its umbrella a suite of behaviors common to a number of species. There will still be interesting philosophical questions about how exactly to understand animal morality, in light of categories and concepts that are central to our understanding of human morality, such as agency, conscience, and impartial judgment. We'll return to some of these concerns in chapter 6.

For now, we want to remind our readers that our focus in *Wild Justice* is moral behavior in social mammals, and we want to assume for now that our definition of morality only applies to the animals under discussion. It is, of course, inevitable that comparative questions will arise, and we do in fact argue that our very general definition of morality can apply equally to human and nonhuman animals, and that it describes essentially the same phenomenon in both. Morality, as a suite of behavioral patterns, is a broadly adaptive strategy. But our focus now is on animals.

Although ancient and contemporary writings in moral philosophy contain a treasure trove of interesting insights, we have found some especially pertinent contributions to the question of animal morality from those in the field who take a relatively "empirical" approach to understanding human morality and the nature of animals. Recently a number of moral philosophers have begun engaging in dialogue with cognitive scientists, moral psychologists, and neuroscientists in an effort to develop a kind of science of morality or at the least to take seriously the implications of science for the philosophical discussion of morality. And a number of philosophers interested in animals have interacted more than casually with ethologists and biologists, and have even begun observing animals firsthand.

The writings of philosophers who challenge stereotypes about animals and who seek to understand and perhaps alter our relationship to animals are, of course, also relevant to a discussion about wild justice. The notion that animals have morality could revolutionize our ideas about who animals are and how we should properly and responsibly relate to them.

We've given you an overview of the broad interdisciplinary scope of research into animal morality. Animal morality sits at the confluence of diverse streams of research, from ethology to neuroscience to philosophy. Now we want to turn to a few particular points about methodology. There are a number of challenges to studying animal minds and emotions and we would like to point out in advance some of the more contentious aspects of our work to preempt some potential objections to, and questions about, the data we present.

EVIDENCE: HOW MUCH IS ENOUGH?

Those skeptical of our work may object that although the available data are suggestive, there is simply not enough there to make a watertight case for animal morality. Indeed, there are gaps in how well scientists understand the social, emotional, and cognitive lives of animals. The longstanding prejudice that animals don't feel or think has meant that investigation into these aspects of animal lives has lagged behind other areas of research in ethology and biology. However, this tide is turning now, and there's considerable interest in exploring the rich inner lives of animals and in trying to understand how animals flourish together in complex societies. There is certainly a great deal more work to be done, and many aspects of the lives of animals will probably always remain a mystery. However, this does not mean that we are unable to make strong and reliable assertions about animal minds and what's going on in them.

There is a preponderance of evidence suggesting that social mammals exhibit a suite of moral behaviors. And new research will almost certainly bolster our case. The research we present is not in and of itself controversial except in rare cases, which we're careful to note. Using the label "morality" is. It is worth reminding ourselves (and the skeptics) that applying the label "morality" to a suite of observed behaviors is a philosophical move, as much as a scientific one. And philosophical objections to this move should not be disguised as scientific objections. Skeptics need to be careful not to confuse or conflate the two.

EMPATHIC ETHOLOGY: CLOUDY OR CLEAR?

After spending considerable time in the field with animals, researchers almost inevitably develop a sense of closeness and even love for the animals they study. Number crunchers may consider this emotional investment in the object of research a confounding factor, likely to somehow cloud the cold, objective vision that scientists should try to have toward the object of study. But in reality, the sense of attachment that allows a scientist to have empathy for the object of study so that the object really becomes a subject allows the scientist intuition and insight that may otherwise remain untapped. Many things about an animal come to light only once we see them for what they are—subjects of their own lives. Jane Goodall broke scientific convention by naming her Gombe chimpanzees Flo and Fifi and David Greybeard, rather than simply referring to them as numbers. And her long-term research on chimpanzees has clearly contributed a great deal to our understanding of these animals and has led to an incredible amount of new research. Consider, too, the reflections of George Schaller, one of the world's preeminent field biologists: "Without emotion you have a dead study. How can you possibly sit for months and look at something you don't particularly like, that you see simply as an object? You're dealing with individual beings who have their own feelings, desires, and fears. To understand them is very difficult and you cannot do it unless you try to have some emotional contact and intuition. Some scientists will say they are wholly objective, but I think that's impossible." Schaller was asked what it was like to stare into the eyes of a gorilla. "I felt a very definite kinship. You're looking at another being that is built like you, that you know is a close relative. You can see and interpret the expression on their faces. In other words, you have empathy with what they're doing. To try to know what an animal is thinking is impossible at this stage in our knowledge of species, but you can interpret their responses on the basis of your own. Besides, they're beautiful, they're individuals. You can recognize all of them by their faces."

Although not universally true, work in the field that seeks to model classical ethology seems to connect with empathy for and love for the animals, seeing them as subjects, not objects. An animal isolated in a lab cage is made into an object for our study. In their natural setting, they're subjects of their own lives, living with their own families and within their own societies. We have the privilege to watch and take notes.

In addition to identifying with the animals we study, we also need to spend a good deal of time with them. Jane Goodall started out with about six months' worth of funding to study chimpanzees at the Gombe Stream Game Reserve but her preliminary findings were so significant that Louis Leakey, who had hired her in the first place, was able to secure funding for more time in the field. Fifty years later, data are still being collected on chimpanzees at Gombe, making this the longest continuous research on animals in a specific location. Since the lifespan of chimpanzees is between about forty and fifty years, Goodall has been there just long enough to witness a full generational cycle. She's been able to observe the entire reproductive and social life of the matriarch Flo, has watched Figan and Freud come into this world, become alpha males, and pass on into old age. She's gotten to know each chimpanzee, and can describe each one's personality and behavioral quirks as if they were close personal friends. It takes this kind of long-term, "immersion" research to collect the data needed to really understand how animals live together as societies, and be able to recognize individual variation in behavior.

Unfortunately, long-term behavioral studies of the sort Goodall and Schaller have championed are decreasing, and are being replaced by short-term studies. Many researchers want to know what animals do, and they want to know it now because knowledge of what animals do in different situations is essential both for making sense of and assessing the relevance of the results of studies of the neural or hormonal bases of behavior. In addition, funding agencies often don't offer up enough money to guarantee that a project will go on for the long term because results are what drive further funding, and frequently there are lapses in the generation of data because of uncontrolled situations in the field, changes in social groups, variations in food supply, or human presence that influences the behavior of the animals being studied and the quality and quantity of information that can be gathered.

Marc is often asked to give a quick summary of his years of research on wild coyotes or on social play behavior so that colleagues can fit in what he discovered with what they're learning about the neural bases of social behavior. But what's missing is an appreciation for the variability that is shown even by members of the same species and for how behavioral flexibility is central to developing theories of social evolution, including the evolution of moral behavior. Scientists often churn out papers based on months, weeks, or days, rather than years or even decades of work. We have a speedy onslaught of neural and molecular data, but the much-needed behavioral data take much longer to collect. It requires patience and the dedication of a lifetime. Results cannot be forced. To get a feel for variability in individual behavior, one needs to watch many, many animals over a long time. An understanding of the larger behavioral context within which individual behavior takes place is essential. And individuals need to be observed under conditions that are as close as possible to those in which they have evolved. As a result of long-term studies of known (and named) individual animals, researchers such as Jane Goodall and George Schaller have learned about the nuances of social behavior and have developed a feeling for the animals, both of which are essential for learning more about what variables underlie social and emotional intelligence.

NARRATIVE ETHOLOGY: STORIES ANIMALS TELL AND WHAT THEY MEAN

We often use stories to make a point or raise a question about animal morality. For example, we'll tell you about an elephant named Babyl who is treated with empathy by her herdmates, and about a chimpanzee named Knuckles whose troopmates make all sorts of behavioral modifications to accommodate his cerebral palsy. Although stories are appealing to many people, some researchers view tales of this sort as nothing more than just-so stories. It's true that anecdotes provide a kind of data that is qualitatively different from the hard numbers of empirical studies, and they cannot substitute for rigorous scientific research. But the use of stories, or "narrative ethology," is an important part of the science of animal behavior. To wit, Lucy Bates and Richard Byrne, working at the University of St. Andrews in Scotland, have recently outlined a formal method for using anecdotes to study animal cognition and have shown them to be extremely useful for learning about the cognitive capacities of elephants, deception in primates, and teaching in animals.

A narrative (from the Latin *narrere*, "to recount," related to *gnarus*, "knowing") is a story, or construction of observed reality, which through its telling gives an event meaning. Narrative is an act of interpretation. Seasoned ethologists often find that numbers and graphs don't do justice to the nuances and beauty of animal behavior. Instead, they often find themselves telling stories from the field to make a point or raise a question. Stories can stimulate thought, activate the imagination of scientists, lead to new questions, represent anomalies, and challenge conventions of thought. Sometimes stories are about surprising, isolated events that challenge the scientific establishment's standing assumptions. The story of Nasty Nick, for example, raises the question in Sapolsky's mind and ours of whether animals can be cruel or mean. And sometimes a single event will elicit competing narratives. Ethologists will disagree over what the events mean, as in the case of Binti Jua.

Narrative ethology, which is practiced by ethologists and other researchers, is not the same as "animal stories" that proliferate on the Web or are told by folks standing around at the dog park. Narrative from seasoned ethologists provides interpretation informed by their knowledge about a particular species and its behavior, and their attention to context and individual peculiarities. The stories included in this book (with the exceptions of Libby leading Cashew, the Tasmanian dog who shares his meal, and the "mice in the sink") are all instances of narrative ethology. The stories about elephants and whales being empathic, wolves playing fair, and chimpanzees showing kindness come from seasoned ethologists and biologists who have devoted years to studying the behavior of these particular species. We believe their observations, their "hard data," and their stories all contain valuable insights.

MAKING SENSE OF WHAT WE SEE

It's important to be able to translate observed behavior into scientific language, but this is a tricky business. For example, we will see in the chapter on cooperation that it can be very difficult to tell whether an observed behavior, such as grooming or group hunts, should really be labeled cooperation, and if so exactly what form of cooperation a given behavior might represent. For this reason, scientists are hesitant to apply language that seems to imply too much about an animal's behavior. The convention in biology and ethology is thus conservative in using language such as *empathy, trust, altruism, cooperation,* or *fairness*. For each particular behavior, whether cooperation, altruism, empathy, or fairness, we follow the convention in ethology of being cautious in applying labels. Where we move beyond convention is this: we argue that equitable, altruistic, cooperative, and empathic behaviors taken together represent a system of morality that functions in certain societies of animals, just as it functions in societies of humans.

THE USE OF ANALOGY: LOOKING FOR SIMILARITIES AND DIFFERENCES ACROSS SPECIES

Ethologists and other scientists often make their arguments using analogies. To reason analogically is to make an inference that if things are similar in some respects, they may also be similar in others. Ethologists, for example, compare humans and other animals and look for similarities (and differences) in any number of features, including brain structure, hormones, physiology, anatomy, and genetics, as well as behavior, facial expressions, vocalizations, and so on. They look at parallels across different species and among different individuals of the same species. We are arguing by analogy when we claim that humans have moral emotions that are tied to certain brain structures, and because animals also have very similar brain structures they may also be experiencing similar emotions. Indeed, the brains of many species show similar neural organization in some of the areas involved with emotions. Researchers have recently discovered an area in the brain called the caudate nucleus that is active when humans are making decisions based on trust. Neuroscientist Reed Montague notes that the caudate nucleus likely receives or computes information about the fairness of a social partner's decision and the intention to repay that decision with trust. There's reason to believe, using analogical inference, that an area of the brain devoted to trust will also be found in animal brains. Arguments from analogy are compelling because of evolutionary continuity among diverse animal species, including humans.

While stressing the importance of evolutionary continuity on the one hand, we need on the other hand to keep the principle of uniqueness at the forefront of our attention. Because animal research has for decades been performed in the service of human needs and desires, there is a habitual inclination to generalize to humans from what we learn about animals. Yet this habit of mind can lead to loose and sloppy science. Each species is unique, and even among a given species there will be individual variation. We cannot safely generalize in the realm of morality from animal behavior to human behavior or, while we're at it, from human behavior to animal behavior. This is why we constantly repeat the mantra "morality is species-specific." Continuity is not sameness. Developmental psychologist Jerome Kagan warns in *Three Seductive Ideas* against the tendency of scientists and lay people to make broad generalizations about abstract psychological processes such as fear, consciousness, or intelligence. None of these terms, he argues, refers to a well-identified and singular property, but rather (much too loosely) to a whole range of processes or behaviors. We must work hard to take apart and distinguish the range and specifics of these phenomena. Furthermore, something like intelligence can only be properly understood with reference to the particulars of age, gender, social context, and, of course, species. In a similar way, "morality" does not refer to a unitary competence, but rather to a whole cluster of related behavior patterns that must be explored with careful attention to particulars of species, age, gender, and social context. Kagan notes that "there is no large body of impeccable, interrelated facts that can be arranged into logically powerful arguments" surrounding morality. The scientific investigation of morality, in humans and nonhumans alike, is in its infancy.

Science relies heavily on inference, and animal-to-human inference has been a cornerstone of biological and biomedical research for centuries. Researchers have developed countless animal models from which they infer effects of pharmaceutical or surgical interventions in human patients. The "dog labs" that have long been one of the core educational instruments of many medical schools teach about the physiology of the human heart by having students look at the hearts of living dogs, the assumption being that there is enough similarity to make this a valuable teaching exercise—that animal to human inference is solid and sound. Yet there has long been a prejudice against human-to-animal inference, which is often labeled anthropomorphism, and is considered deeply suspicious.

Some scientists complain that using "human" language to describe the behavior of animals is anthropomorphizing, or attributing human characteristics to nonhuman beings (literally, giving human *antrhōpos* shape *morphē*). This, like the antipathy toward anecdote, is a prejudice that science needs to get over. The term *anthropomorphism* gets thrown about in science, typically as a criticism of someone's work, as if *anthropomorphism* were a synonym for *sloppy*. Ironically, however, critics' use of this term is often so loose and imprecise that it's nothing more than a kind of vague insult. And, speaking of sloppy science, as Marc points out in his book *The Emotional Lives of Animals*, how ironic it is that critics of anthropomorphism get uneasy when someone claims, for example, that a captive animal is unhappy, yet they fail to recognize that they, too, are being anthropomorphic when they counter, "Oh, no, you're wrong, she's happy."

It is particularly when emotions are attributed to animals that charges of anthropomorphism are leveled. This is a result of dogma lagging behind science. There still remain a few researchers, even some ethologists, who have trouble with the idea that animals have emotions. But this is a philosophical trouble they're having, not a scientific one. They may be uneasy with the idea that animals are that much like humans, or humans that much like animals. What scientists who study animal emotions such as fear, joy, and jealousy are doing is not anthropomorphism.

It is science. It's using concepts that have relatively clear meaning within science and exploring how these concepts are expressed in animals.

There's nothing unscientific about using the same terms to refer to animals and humans, particularly when we're arguing that the same phenomenon is present across species. Empathy is empathy. It may be expressed and felt differently in different species and even among individuals of the same species. Yet, across species in which empathy has evolved, there can be little doubt that it emerges out of the same neural architecture and is displayed in similar social contexts, such as when a mouse empathizes with another mouse in pain or an elephant consoles a friend in distress. Instead of using the term *empathy*, we could offer alternative descriptions involving neural circuitry, muscle movements, body temperature, EEGs, and genetic signaling, for example, but these are neither more interesting nor more accurate. Such sanitized and supposedly parsimonious descriptions exclude the social context that is so very important in discussions of animal emotions and animal morality.

Evolutionary continuity suggests a fluid movement in both directions, from animals to humans, and from humans to animals. It makes sense to have symmetry in our comparisons, particularly when it comes to research into animal feelings, mental states, and moral behaviors. It isn't that we set out looking for humanlike traits in animals and hope to find some. Rather, we set out to understand what animals are like, and use the language and concepts that come closest to describing what we see. Consider the words of Sarita Siegel: "The more time I spent with orangutans, the more firmly I was convinced that great apes possess intentionality, self-awareness, complex modes of communication, a theory of mind, a sense of humor, and a need for emotional support, as well as many other human-like traits. For these reasons I felt anthropomorphic analogy and anecdotes were relevant and helpful."

Canadian biologist Hal Whitehead, who is recognized by his colleagues as one of the world's leading whale researchers, wrote:

In the late 1990s two remarkable novels were published: *White as the Waves*, *a retelling of Moby Dick from the perspective of the whale* . . . and *The White Bone*, about the destruction of elephant society as seen by elephants . . . Both novels use what is known of the biology and social lives of their subject species to build pictures of elaborate societies, cultures, and cognitive abilities. Their females are concerned with religion and environment as well as the survival of calves: their males inhabit a rich social and ecological fabric of which mating is only a small part. A reductionist might class these portraits with Winnie-the-Pooh as fantasies on the lives of animals. But for me they ring true, and may well come closer to the natures of these animals than the coarse numerical abstractions that come from my own scientific observations.

Renowned paleobiologist Stephen Jay Gould also noted: "Yes, we are human and cannot avoid the language and knowledge of our own emotional experience when we describe a strikingly similar reaction observed in another species." Anthropomorphism endures because it is a necessity, but it also must be done carefully, consciously, empathetically, and from the point of view of the animal, always asking, "What is it like to be that individual?" We must make every attempt to maintain the animal's point of view. We must repeatedly ask, "What is that individual's experience?" Claims that anthropomorphism has no place in science or that anthropomorphic predictions and explanations are less accurate than mechanistic or reductionist explanations are not supported by any data. Careful anthropomorphism is alive and well, as it should be.

No matter what we call it, we all agree that animals and humans share many traits, including emotions. We're not inserting something human into animals, but we're identifying commonalities and then using human language to communicate what we observe. In an interview in *Salon* magazine, primatologist Robert Sapolsky remarks, "Do I get grief for the fact that in communicating, say, about the baboons I'm doing so much anthropomorphizing? One hopes that the parts that are blatantly ridiculous will be perceived as such. I've nonetheless been stunned by some of my more humorless colleagues to see that they were not capable of recognizing that. The broader answer, though, is I'm not anthropomorphizing. Part of the challenge in understanding the behavior of a species is that they look like us for a reason. That's not projecting human values. That's primatizing the generalities that we share with them."

When we anthropomorphize, we're just doing what comes naturally. Among early humans, anthropomorphizing may have allowed hunters to better predict the behavior of the animals they hunted, and it's very useful for learning more about beastly passions today. It may very well be that the seemingly natural human urge to impart emotions onto animals—far from obscuring the "true" nature of animals—actually reflects a very accurate way of knowing. Alexandra Horowitz and Marc have shown that animals continually provide prompts for anthropomorphizing and it's expected that we would use these to describe and explain their behavior, intentions, beliefs, and emotional states.

Wikipedia contains an entry for *anthropomorphobia*—the fear or hatred of acknowledging in nonhuman animals the qualities that we want to consider distinctively human. Assigning moral behaviors such as loyalty and compassion to animals will certainly evoke this phobic response in some people. We hope that after reading our book their fears will be allayed.

READING THE INNER WOLF

Critics are often quick to exclaim that the emotional lives of animals are too private or hidden to make much sense of. And, surely, animals will always have their secrets. Yet their inner emotional and moral lives are surprisingly public and transparent. Just look at them, listen to them, and if you dare, smell the odors that pour out when they interact with friends and foes. Look at their faces, tails, bodies, gaits, and most importantly their eyes. What we see on the outside tells us a lot about what's happening inside animals' heads and hearts.

People around the world, including researchers, readily recognize expressions of emotions and show remarkably high agreement when asked what they infer about what an individual animal is feeling based on their observations. Behavioral scientists Françoise Wemelsfelder and Alistair Lawrence tested the hypothesis that every observer, whether trained in animal behavior or not, can make a meaningful assessment of an animal's behavior. Trained and untrained observers displayed a high level of agreement about what emotions an animal is feeling. These results constitute important data and suggest that the problem of never being able to enter into the subjective experience of another, what philosophers call "the problem of other minds," isn't really so serious after all.

There is, of course, no complete or final resolution to the problem of other minds. No matter how neatly we slice apart the brain and peer at different bits under a microscope, we will never know *exactly* what it is like to be a wolf. So, when Lupey, a male wolf, invites Herman, another male wolf, to play, we can only infer that Lupey wants to play and that Herman knows this and also wants to play. However, armed with detailed knowledge about social play behavior in wolves, we're able to make extremely accurate predictions about what follows when Lupey solicits play from Herman. In wolves and other animals, their public displays reveal a lot about what is happening inside their heads and there really isn't all that much guesswork.

Let's get to the heart of the matter. The problem of other minds is *not* an impediment to understanding how animals feel and think. Why not? Well, first of all, animal minds aren't all that inaccessible or private, as cognitive ethology and social neuroscience make abundantly clear. They are, actually, rather public. We already know a lot about animal minds, and we're discovering more and more each day. Second, and perhaps even more important, we are ourselves animals and our experiences of pain, joy, envy, compassion, and love are probably very much akin to these same emotional states in other animals. Data suggest that there is enough continuity in physiology and psychology to safely infer significant experiential common ground. And finally, we must remember that human minds are private, too. We can never crawl inside the skin or brain of another person and truly know their subjective experiences. Yet this doesn't stop us from understanding and reacting to their thoughts or emotions, most of the time quite accurately and without conscious effort. The so-called privacy-of-mind problem is overused and is little more than a poor excuse for ignoring much ongoing research and retaining the status quo in our treatment of animals.

ANIMAL EMOTIONS AND FELLOW FEELINGS

The emotional lives of animals have been the underbelly of animal behavior research. It has been assumed either that animals don't experience emotion, or that their emotional lives are so simple as to be uninteresting. Until quite recently, even, emotions in animals were catalogued as simple behavioral responses, reducible to chemical changes in the brain or body. Fear, for example, was described as just a physiological event—the "flight or fight" response describes the release of catechol-amine hormones, leading to constriction of blood vessels, acceleration of the heart and lung functions, and so forth. Well, human emotions can be reduced in the same way, but most people recognize that this is an impoverished picture of what it means to have a feeling such as fear, and that fear has many faces. Fortunately, all this is changing, and we now know that the emotional lives of animals are every bit as rich as our own. There's a lot of interest in animal emotion and lots of new research (see, for example, Marc's *The Emotional Lives of Animals* and Jonathan Bal-combe's *Pleasurable Kingdom*). The tendency to focus on "negative" emo-tions such as pain, fear, and aggression has given way to an increased interest in "positive" emotions such as love, joy, and pleasure, and to complex emotional experiences such as empathy, grief, and forgiveness. The emotional lives of animals are at the heart of animal morality, and new research into animal emotions will certainly push the development of this young science.

FOUNDATIONS OF ANIMAL MORALITY: SOCIALITY AND INTELLIGENCE

Our general hypothesis is that the complexity of moral behaviors and the development of what we call moral intelligence in animal species depend on both sociality and intelligence. Morality is an evolutionary adapta-tion to social living. Many of us tend to think of animals as individual units—the dog lying underneath my desk, or the squirrel scurrying along the fence toward my bird feeder. But for animals, as for humans, life is really all about social relationships. As Animal Planet's popular show *Meerkat Manor* suggests, the lives of animals are every bit as much a soap opera as the lives of humans. Animals form friendships, are caught ly-ing or stealing and lose face in the community, they flirt, their sexual advances are sometimes embraced and sometimes rejected, they fight and make up, they love, and they experience loss. There are saints and sinners, bad apples and good citizens.

Sociality is the tendency of an animal to associate with others in long-lasting social groups. Of the myriad species on the planet only a small fraction has achieved a high level of social complexity. In *Sociobiology*

FIGURE 2. Polar bears showing affection for one another in Hudson Bay, Manitoba, Canada. Courtesy of Thomas D. Mangelsen/Images of Nature.

E. O. Wilson described four groups of creatures that in his view represent the pinnacles of social evolution, namely, the colonial microorganisms and invertebrates (such as slime molds and corals), the social insects (bees, wasps, ants), the highly social vertebrates, and humans. Our principle focus is the social vertebrates, specifically the social mammals, though we make frequent reference to humans. The evolution of morality is, of course, a small piece of the larger picture of the evolution of sociality, and this, as a broad evolutionary phenomenon, is an important backdrop for our discussion.

Although we have good data to support the claim that a small range of social mammals has moral behavior, there really isn't enough information to make hard and fast conclusions about other species. And even if other forms of life lack morality, we may still have much to learn by studying their diverse forms of sociality. James Costa's *The Other Insect Societies*, for example, challenges the study of insect sociality to extend beyond its single paradigm of sociality shaped by kin selection, as we see in the eusocial arrangements of ants, bees, and wasps. His work points to a diversity of social arrangements, and suggests that there may be many evolutionary paths to sociality, not all of which involve kin selection. Likewise, we may find that mammalian sociality, if studied with an open mind, cannot be adequately understood within the currently prevailing paradigms, and we likely will be pushed into a richer theoretical framework.

INDIVIDUALS AND GROUPS: THE GIVE AND TAKE OF SOCIAL LIVING

Just about all mammals display some level of sociality, enough at the very least to mate and perhaps to tend young. But social mammals take sociality to a different level. They're highly interactive to the point that individuals live together in recognizable societies and form enduring relationships with other members of their group. A relationship involves repeated encounters over time, where each interaction is affected by memory of past interactions and expectations about future interactions. Relationships are patterns of coordination among individual animals; an animal will act and feel in reference to the actions or feelings of another. Relationships in turn take place in the context of larger social groupings (families, clans, and societies).

In many social groups individuals establish social hierarchies and develop and maintain tight bonds that help to regulate social behavior. Individuals coordinate their behavior—some mate, some hunt, some defend resources, and some accept subordinate status—to achieve common goals and to maintain social cohesion. As Robert Sussman and Audrey Chapman note in *The Origins of Sociality*, group-living animals must give up part of their individual freedoms in order to be a functioning part of the group. Sociality, then, refers to "the compromises that individuals make, the mechanisms they use, and the means by which they maintain these social groups."

Daniel Goleman suggests, in *Social Intelligence*, that people who wind up running Fortune 500 companies have excelled in business not because they are school-smart, but because of their social intelligence, their ability to read people, to form friendships and alliances, and to anticipate and respond appropriately to the desires of others. For other highly social animals, too, social savvy can be an important factor in survival and reproductive success. For example, Robert Sapolsky studied how social life for a baboon society influenced levels of the stress hormone cortisol in individual animals. Social stress is a large part of baboon life. There is constant maneuvering over rank, for example, with higher-ranking individuals intimidating and harassing lower-ranking individuals. This can be very stressful for a lower-ranking animal. Sapolsky went on to show that stress can have health consequences for the animals, including elevated blood pressure. Females who are stressed out have more trouble producing healthy offspring. He also found that individual baboons vary considerably in their ability to handle stress, and that those with the most stable social connections seem to handle stress the best. Males who spent more time grooming and being groomed, and playing with babies, had lower levels of stress hormones. This relationship between social connection, stress, and health has been observed in humans, too.

Animals have various means of maintaining social order, including direct negotiation, third-party mediation, and reconciliation, all manifestations of what Frans de Waal calls community concern or "the stake each individual has in promoting those characteristics of the community or group that increase the benefits derived from living in it by that individual or its kin." Community concern begins to look suspiciously like morality: those behaviors (deceiving, cheating) that tear the social fabric are "wrong" and those that create the kind of community in which individuals thrive are "right."

INTELLIGENCE, BEHAVIORAL FLEXIBILITY, AND MORALITY: WHAT ARE THE CONNECTIONS?

Animals with complex moral behaviors are not only highly social, but also highly intelligent. Ethologists tend to define intelligence as an aggregate of special abilities that have evolved in response to specific environments, and that allow individuals to adapt and to be behaviorally flexible in varying circumstances. This is obviously a loose definition, but the looseness is deliberate. Intelligence is not a single capacity or ability, nor is it something that can be easily or meaningfully compared between or even within a species. It isn't especially meaningful, for example, to ask if cats are more intelligent than dogs. Cats do what they need to do to be cats and dogs do what they need to do to be dogs. While it might be useful to compare members of the same species in terms of how smart they are, this too might be fraught with misleading inferences. If Fido, a dog, learns where his food is faster than his canine buddy Herman, is Fido smarter? Perhaps, but what if Herman learns to avoid cars more rapidly than Fido? Is Herman more intelligent? Are midwife bats who help another bat give birth more intelligent than nonmidwife bats because the former recognizes that another female is having a difficult labor? Who knows? And what about cultural variations in the manufacture and use of tools by chimpanzees? Are chimpanzees who use tools more intelligent than chimpanzees who don't? It's unlikely that they are. Specific circumstances have led to the use of tools and it's likely that all chimpanzees with normal chimpanzee brains would, in the right context, display an innovative ability to make and to use tools. Along these lines, Gerhard Roth and Ursula Dicke argue that intelligence has evolved independently in different classes of vertebrates, which speaks against an "orthogenetic" view of intelligence in which there is a single evolutionary trajectory culminating in Homo sapiens.

We defined intelligence as how well an individual adapts to his or her particular environment. There is no general intelligence. Intelligence is not a universal and measurable entity. Jerome Kagan writes, "The defenders of [general intelligence] . . . , like those who believe in one fear state or one type of consciousness, fail to appreciate that organs and physiological systems develop independently. No single general factor can represent the growth rates of diverse classes of cells, tissues, and organs in animals or humans. The description 'intelligent' is frequently found in sentences that are indifferent to the age and background of the person (or sometimes the animal species) or the evidential basis for the assignment." Intelligence is context-specific. And to reiterate, cross-species comparisons, or even within-species comparisons, are fraught with difficulty.

Intelligence is often equated with cognitive complexity, with, for example, causal reasoning, flexibility, imagination, prospection, and memory. These are, indeed, important aspects of intelligence. But they're only part of the picture. Harvard University researcher Howard Gardner deepened our understanding of human intelligence by suggesting that there are multiple intelligences. Human intelligence has at least six facets: linguistic, musical, logical-mathematical, spatial, bodily-kinesthetic, and personal. Animals too have multiple intelligences, though the list will look different for each species.

THE SOCIAL INTELLIGENCE HYPOTHESIS

Early speculation by primatologist Alison Jolly, and later by psychologist Nicholas Humphrey, about the seemingly unique complexity of primate social interactions led to several intriguing questions: Is there a connection between the large size of the primate brain and the complexity of primate social life? How closely are sociality and intelligence linked? One of the most provocative recent ideas in the study of behavior, the "social intelligence hypothesis" (or SIH, sometimes also called the Machiavellian intelligence hypothesis), arose in answer to these questions. The basic idea behind the social intelligence hypothesis is that development of social skills drove the evolution of intelligence, at least among primates.

Animals living in groups may do better (as do the groups themselves) when individuals can manipulate social information and social relationships, when they can keep track of who has helped them, who is untrustworthy, who is allied with whom, and so forth. And clueing in and keeping track of such nuanced information requires a flexible, complex, and relatively large brain. Variations on the social intelligence hypothesis have focused on several aspects of social behavior that seem to require advanced cognitive skills including the formation of alliances and coalitions, the use of deception, and the transmission or teaching of novel behaviors.

A related hypothesis is that brain size is correlated with group size: the larger the social group an animal has to manage, the more brain power is needed (and brain power, in this view, correlates with brain size). A number of studies in social mammals have shown a correlation between mean group size and neocortex volume: the larger the social group, the larger the neocortex (the part of the brain involved in higher-order processing of social information). Various primate species show this correlation, as do bats, carnivores, and toothed whales. Yet correlation does not imply causation, and some of the conclusions about the relationship between social-group size and brain size remain tentative.

What seems to be emerging now is what might be called a "multifactorial hypothesis." It may be that all of the competing variations of the SIH have an element of truth and that social complexity and/or group size are only one or two of many factors that have influenced the evolution of intelligence. SIH may provide only a partial answer to why greater intelligence evolved in certain animals and not others. Alternatives to the SIH, such as the "foraging hypothesis"—the idea that foraging strategies (such as whether an animal ate leaves or fruit) provided the selection pressure that led to increased intelligence—may provide complementary, rather than contradictory, evolutionary explanations.

Our hypothesis about moral behavior in animals does not require something like the social intelligence hypothesis. But the SIH is very suggestive and the research focused on trying to sort out the relationships between sociality and intelligence is pertinent to our project. We can gain particularly important insights from criticisms of SIH. Indeed, SIH does appear to have important limitations and counter-examples. Perhaps the most significant limitation of SIH is that it has been developed in relation to primates and has primarily relied on behavioral studies of primates. Even if SIH provides a compelling hypothesis about primate intelligence, it may or may not apply to other species of animal. Many counterexamples can be found. For instance, members of the bear family, typically solitary animals, have greater relative brain and neocortex size than highly gregarious carnivores. Scrub jays have episodic memory and can plan for the future, two highly advanced cognitive skills, yet as birds go scrub jays are relatively unsocial.

Hyena expert Kay Holekamp argues that there is a great deal of work still to be done in understanding the social intelligence hypothesis and that we should be cautious about brain-size-to-sociality correlations. For example, the brain size of mammalian carnivores and their ungulate prey are known to co-vary over geological time—as prey brain size increased so did that of their carnivorous predators. Holekamp points out that these changes occurred both in solitary and gregarious carnivores, a trend that is not predicted by the social intelligence hypothesis. Selection pressures during evolution are rarely singular. Selection pressures associated with sociality interact with other selection pressures, such as the demands of a complex environment. While the social intelligence hypothesis makes interesting predictions, many of which are supported, Holekamp correctly notes that in the future we need models involving many different variables.

We also need to extend investigations into the sociality-intelligence connection by careful studies on nonprimates. A recent controversy over the intelligence of dolphins provides a fascinating case study on nonprimate sociality, and suggests that much can be learned by going beyond the confines of the primate paradigm. In 2006, Paul Manger made the controversial suggestion that water temperature and not social complexity was the major selection pressure that led to the development of large brains in cetaceans, and that dolphin brains are big because they have a lot of thermal padding. In response to Manger's paper, dolphin expert Lori Marino and her colleagues did a meticulous review of existing data on dolphin sociality and dolphin intelligence. They argued that the social intelligence hypothesis actually fits the dolphin data quite well. Bottlenose dolphins live in very complex societies, with intricate systems of communication, collaboration, and cooperation, as well as competition. They form simple alliances, higher-level alliances, as well as long-term bonds. Dusty dolphins are known to cooperate with one another to keep a large ball of anchovies upwards of one hundred feet in diameter intact so that the dolphins can all feed. There is even evidence for individual role-taking in dolphin societies, to facilitate cooperative relationships and decision-making processes, all of which support the hypothesis that dolphins have advanced cognitive skills.

There's still one more piece to the sociality-intelligence question: how does moral behavior link with the complexity (and/or size) of social organization and with social intelligence? This question is as yet unexplored, but it will be a fruitful avenue for further research. Our proposal is that the development of moral behaviors is closely linked both with complex sociality and intelligence: the more complex a species' social network, the more complex are individual repertoires of moral behavior, and the more complex is moral behavior, the more socially intelligent individuals are.

GREATER SOCIAL COMPLEXITY =
MORE NUANCED MORAL BEHAVIOR

Our hypothesis is that greater social complexity is linked with more complex and nuanced moral behaviors. Does this mean, then, that solitary animals such as tigers and wolverines will lack these behaviors? Not necessarily. Sociality and solitariness are not opposites, but rather two points on a continuum. There are few if any truly solitary individuals, because most individuals interact with others of their own or other species. Consider the domestic cat, a paragon of solitary self-sufficient existence. Are cats truly solitary? Of course not. As the research of ethologist Paul Leyhausen showed, cats are extremely sensitive to, and interested in, the olfactory signals of other cats, which, in turn, are emitted not randomly, but with the express intention of communicating information to other cats about territory and gender. These count as social interactions. There also is variation within species. Wolves typically live in packs, but there are also lone wolves.

In contrast to wolves, wolverines are highly solitary, and probably have evolved few of the mechanisms we're discussing. But we wouldn't say that they lack moral behaviors. Most likely, they have relatively little use for such behaviors. So if you want to examine moral behavior in animals, wolverines would not be the best animals to study. You should look instead at highly social animals—wolves, hyenas, bonobos, meerkats—in which there is a variety of complex social interactions.

MORALITY AS A UNIFYING CONCEPT:
TYING IT ALL TOGETHER

During the past decade there has been marked increase in attention to prosocial behaviors in animals and to the recognition that the lives of animals are not just shaped by competition and conflict. We now know that animals have a large repertoire of prosocial behaviors, and even some building blocks of morality. But the various pieces of the puzzle (empathy, cooperation, fairness) have not yet been pieced together into a coherent whole.

The concept of animal morality encourages a unified research agenda. An exploration of moral behavior in animals allows a number of seemingly distinct research agendas in ethology—research on animal emotions, animal cognition, and diverse behavior patterns such as play, cooperation, altruism, fairness, and empathy—to coalesce into a coherent whole. Animal morality also unifies threads of research from diverse disciplines, from ethology, of course, but also from philosophy, neuroscience, psychology, and more. And this is what's so exciting about comparative research into the social behavior of a wide range of animals.

We're using the phrase *animal morality* to refer to the suite of other-regarding behaviors and capacities that nurture and foster social encounters and allow for the flexibility that is needed so that individuals can adapt to different social contexts. This suite of behaviors includes cooperation, empathy, and justice, as well as the social, cognitive, and emotional intelligences that make such behaviors possible. Let's now turn our attention to exploring in detail this suite of moral behaviors.

3 COOPERATION
RECIPROCATING RATS AND
BACK-SCRATCHING BABOONS

If you follow science news you will have noticed that cooperation among animals has become a hot topic in the popular press. For example, in late 2007 the science media widely reported a study by zoologists Claudia Rutte and Michael Taborsky suggesting that rats display what they call "generalized reciprocity," providing help to an unfamiliar and unrelated individual, based on the rat's own previous experience of having been helped by an unfamiliar rat. Rutte and Taborsky trained rats in a cooperative task of pulling a stick to obtain food for a partner. Rats who had been helped previously by an unknown partner were more likely to help others. Before this research was conducted, generalized reciprocity was thought to be unique to humans and perhaps chimpanzees.

Finding complex cooperative behavior in rats may seem remarkable. But it really isn't surprising. Rutte and Taborsky's research simply adds another piece to a large database centering on cooperation among a diverse range of animals. To give two more examples, Amanda Seed, Nicola Clayton, and Nathan Emery discovered that rooks, a member of the crow family, would team up and cooperate to reach a tray of food that a single bird couldn't access alone. And, zoologists Christine Drea and Laurence Frank discovered that captive spotted hyenas cooperate with one another to acquire food, even without specific training. They observed pairs of adult spotted hyenas cooperating to solve a synchronous rope-pulling task, in order to open a trap door. When the door opened, food dropped to the ground and both hyenas could eat. Drea and Frank also observed that hyenas showed behavioral flexibility while cooperating. In other words, individuals modified their behavior to accommodate various partners, including those with no knowledge of what was required in the task. The animals not only visually monitored their partner's behavior, but also changed leadership roles and switched positions in order to get the food.

The recent deluge of essays and research papers on cooperation shows that the more we look for cooperation in animals the more we discover its presence. And indeed, if you watch animals for any length of time, it's easy to see a good deal of cooperating and plain old getting along. Cooperation can be thought of as the superglue that binds and maintains social ties among animals. In fact, you see much more cooperative and tolerant behavior than teeth, claws, and blood. Even in situations where you might expect to see competing and fighting, such as over a tasty meal, cooperation tends to prevail. Wolves, for example, hunt in long-lasting packs and cooperatively defend their kill from other animals. In most situations, food is distributed so that all group members get what they need, although subordinate individuals may have to wait to indulge until higher-ranking animals have had their fill. Even individuals of different species also sometimes work together. Bernd Heinrich and his students discovered that ravens will lead wolves to an elk carcass. The wolves tear the carcass apart (a job the ravens cannot themselves execute) and feast, and the ravens are then able to eat, too. Marc has observed the same sorts of interactions between ravens and coyotes.

In a 2005 *Scientific American* essay Frans de Waal argued that human tendencies such as reciprocity, division of rewards, and cooperation aren't limited to our species. He wrote, "They probably evolved in other animals for the same reasons they evolved in us—to help individuals take optimal advantage of one another without undermining the shared interests that support group life." De Waal used the example of food sharing in capuchin monkeys and chimpanzees to argue the point: "This reciprocity mechanism requires memory of previous events as well as the coloring of memory such that it induces friendly behavior. In our own species, this coloring process is known as 'gratitude,' and there is no reason to call it something else in chimpanzees."

Cooperation is widespread, but its manifestations among animals are complex and diverse and require a rich set of cognitive and emotional skills. Cooperation is one of the fundamental building blocks of moral behavior. Here we explore a range of cooperative-behavior patterns, and look for those instances of cooperation that might fit within our suite of moral behaviors.

THE STRUGGLE FOR EXISTENCE: BALANCING COMPETITION WITH COOPERATION

Stephen J. Gould continually reminds us that Darwin used the phrase "struggle for existence" metaphorically, and that even Darwin understood that bloody and vicious competition is only one possible mechanism through which individuals might achieve reproductive success. Another possible mechanism was proposed by a contemporary of Darwin, Russian anarchist Peter Kropotkin, in his forward-looking book *Mutual Aid*, published in 1902. Kropotkin suggested that cooperation and mutual aid may also lead to increased fitness, and may more accurately fit our actual observations of animals in nature. Although biologists have largely explored cooperative behavior through the Darwinian lens of competition and an evolutionary arms race, we might wonder what the intellectual history of evolution would look like had Kropotkin's ideas been taken more seriously.

In *Mutual Aid* Kropotkin laments that although he "looked vainly for the keen competition between animals of the same species which the reading of Darwin's work had prepared us to expect . . . facts of real competition and struggle between higher animals of the same species came very seldom under my notice." What he likely saw was a good deal of cooperative behavior, with a smattering of aggressive and competitive behavior now and then. Researchers Robert Sussman, Paul Garber, and James Cheverud looked closely at published data on primate social behavior and what they saw, like Kropotkin, was that the overwhelming majority of social interactions in a variety of primate species were affiliative rather than agonistic. Most of the time these animals are friendly and cooperative with each other. Sussman and his colleagues concluded "friendly, peaceful, coordinated, and cooperative interactions serve a greater role [than agonistic interactions] in alliance formation, friendships, social cohesion, and obtaining access to resources, and have utility outside of combating or ameliorating aggression." Jane Goodall made similar observations during her long research on chimpanzees at the Gombe Stream National Park, and Marc has noted similar patterns among social carnivores. Across species, cooperation and affiliation are the major governing principles of animal sociality.

WHY COOPERATE? WHAT'S WORKING TOGETHER GOOD FOR?

Animals cooperate for many different reasons. They work together to protect themselves, either from other group members or from other animals. For example, female chimpanzees form groups to protect themselves from aggressive males, and large flocks of chaffinches will group together to mob intruders. Animals will also take turns feeding and scanning for predators. For example, related and unrelated meerkats take turns as sentinels, some individuals scanning for predators while others eat. Western evening grosbeaks and numerous other species of birds show similar patterns of trading off feeding and scanning. Other common behavior patterns such as alliance formation, communal nursing and care of young, and grooming others are also examples of cooperation. For example, male dolphins form social groups called "super alliances" to gain access to females, and female rats often nest and nurse young communally, even sharing milk. Primates maintain enduring social bonds by grooming one another in complex social networks, as do ungulates. Sure, there are cheaters, liars, and free riders in all of these cooperative systems, but the rule breakers are the outliers, the exceptions to the norm. Cooperative behaviors are everywhere and act as the superglue of animal societies.

What do we make of this ubiquity of cooperative behavior? Why has cooperation evolved in so many species? Cooperative behavior has always been a puzzle because it doesn't fit the predictions of Darwinian theory, which has sent us looking instead for competition and unfettered aggression. It must be that evolution, though a "competitive" process, doesn't yield only competitive, ruthless, aggressive strategies. Evolution clearly can and does also give rise to strategies of cooperation and nice-ness. Cooperation, in its turn, allows specialization and thus promotes biological diversity. Martin Nowak, director of the Program for Evolutionary Dynamics at Harvard University, has argued that cooperation is one of the three basic principles of evolution, alongside mutation and selection. "Cooperation," Nowak says, "is the secret behind the open-endedness of the evolutionary process. Perhaps the most remarkable aspect of evolution is its ability to generate cooperation in a competitive world."

THE COOPERATION CLUSTER AT A GLANCE

We use "cooperation" as shorthand for an entire suite of behaviors related to helping others and working together with others toward a common goal, and we present data on a wide range of cooperative behaviors—grooming, group hunting, communal care of young, alliance formation, and play—to explore the concepts of cooperation, altruism, and reciprocity. We also give attention to various mechanisms that grease the wheels of cooperation: honesty, trust, punishment and revenge, spite, and the negotiation of conflicts.

Cooperation and its behavioral relatives form an important part of the suite of moral behavior in animals. Still, most instances of cooperation are not "moral" in our sense, because we've limited our moral suite to behaviors in which there is a certain level of cognitive complexity and emotional nuance. It's important to recognize that cooperation is everywhere in nature, and that it is serves to foster relationships and societies in which morality blossoms. We need to examine the larger phenomenon of cooperative behavior in animals, while at the same time trying to identify those cooperative behaviors that can confidently be labeled morality.

Although we draw on examples of cooperation such as grooming, cooperative hunting, and food sharing, we're particularly interested in what lies beneath these behaviors, what cognitive and emotional capacities allow animals to engage in these and other kinds of cooperative social interactions. De Waal argues for a similar point of emphasis: "In discussing what constitutes morality, the actual behavior is less important than the underlying capacities. For example, instead of arguing that food-sharing is a building block of morality, it is rather the capacities thought to underlie food-sharing (e.g., high levels of tolerance, sensitivity to others' needs, reciprocal exchange) that are relevant."

SOME PRELIMINARY CLARIFICATION OF TERMS: BIOLOGICAL VERSUS EVERYDAY JARGON

Yvette Watt, an artist and animal advocate in Hobart, Tasmania, told Marc the following story about two dogs. One is a well-fed and happy canine, the other a sad dog who is always tied to a rope. The happy dog's daily walk takes him by his unfortunate neighbor. One night, the happy dog eats his usual dinner, but saves his meaty bone. The next morning, he carries the meaty bone on his walk, and delivers it to his tethered friend. Lorraine Biggs, who told this story to Yvette, saw the happy dog's behavior as an act of altruism. But is it really? The meaning of altruism in biology is not quite so straightforward, nor is much of the other language used to talk about cooperation among animals.

Certain terms in our cooperation cluster, like *altruism* and *spite*, have particular meaning in biological parlance that diverges from the normal use of the words in everyday conversation. *Altruism*, in our daily language, refers to an unselfish concern for the welfare of others, with emphasis on unselfish. If your motive for helping an elderly person cross the street is that you want to be nominated for citizen of the month, you haven't really acted altruistically. In biology, altruism lacks this moral coloring; there's no accounting for intention or motive. When biologists talk about altruism in nature, they use the language of costs and benefits, which are cashed out as reproductive-fitness consequences. As philosopher Elliott Sober and evolutionary biologist David Sloan Wilson note in their book on the evolution of unselfish behavior, *Unto Others*, "biologists define altruism entirely in terms of survival and reproduction." Altruism refers to a behavior that is costly for the actor (it decreases reproductive fitness) and beneficial for the recipient (it increases reproductive fitness). In biology, "altruistic" does not equal "moral."

We also need to stress the potential ambiguity of the related term "selfish" in discussions of animal morality, where scientific and common meanings can easily get blurred or confused. The concept of "self-ishness" in biology, popularized by Richard Dawkin's influential book *The Selfish Gene*, is amoral; it refers simply to the inclination or "drive" of each gene to promote its own reproductive success. (Genes, as far as we know, do not have intentions.) The evolution of morality, including unselfish behavior, is perfectly consistent with theories of "selfish genes." We just need to remember that explanations about why a behavior evolved and what makes an animal exhibit that behavior right now are distinct. Unfortunately, it is almost impossible to wipe away the moral connotations etched into the word; even scientists seem to forget sometimes. Let's be clear then. Selfish genes and moral animals—and not just apparently moral, but really, truly moral—are entirely comfortable together, as evolutionary phenomena.

To be complete, we need to mention spite, which also has a specific technical meaning in biology. *Spite* refers to behavior in which all individuals pay: the actor incurs a reproductive cost in order to punish the recipient, who also incurs a reproductive cost (for failing to cooperate, or for cheating in some way). Although some animals might feel rancor toward those they punish, *spite* in its technical sense would carry no moral weight. The existence of spite in nonhuman animals is highly questionable, and experts agree that at this point there are no reliable accounts of this phenomenon except for a much-debated report of spite among siblings in a species of parasitic wasp.

Some of our terms have no special meaning in biology, notably *co-operation* and *reciprocity*. There's no special biological definition that is distinct from ordinary usage. Cooperation is behavior in which both parties benefit at the time of the interaction. There is typically no cost to the cooperators, only benefit. Reciprocity is a form of mutual social exchange—you scratch my back and I'll scratch yours. I may incur some cost now, in order to benefit you, with the expectation that you will later incur some cost to benefit me. In reciprocal exchange, the sharing of favors sometimes stretches out in time—you help now so that later you might in turn be helped. Of course, neither cooperation nor reciprocity are generally considered moral virtues in humans, nor is the lack of these behaviors the mark of an evil or bad person (antisocial, perhaps, and not a desirable corporate employee, but not in need of one's Sunday prayers).

This is perhaps why no effort has been made to give these concepts a special meaning in science, for they don't carry much moral weight. It makes our job a bit harder, though, because some concepts we want to "de-moralize," whereas others we want to "moralize."

In the scientific literature cooperation is sometimes taken to be synonymous with altruism, while at other times it is distinguished from altruism and reciprocity as a particular behavioral category. It's hard for us to avoid some of this ambiguity, as we draw on work that uses this whole spectrum of meanings. We have chosen to let the word cooperation function broadly throughout most of this chapter, thus, we take altruism and reciprocity as two specific types of cooperative behavior.

In addition to some ambiguity in how language is used, there is an additional problem in labeling cooperative behavior. Often, it is very difficult to know whether an observed behavior should appropriately be labeled altruism (cost for actor, benefit for recipient) or cooperation (benefit for both). For example, even if we assume that happy dog and sad dog in Yvette Watt's story are not genetically related, we really don't know if happy dog was paying back sad dog for something sad dog had done in the past. This is important to know, because the rigorous definition of altruism suggests that happy dog incurred some cost, and no benefit, by bringing his meaty bone to sad dog. Truth be told, in the vast majority of studies on altruism and cooperation in animals, researchers don't know genetic relationships among individuals, nor is it always easy to know whether an animal has incurred some cost or benefit in terms of the loss or gain in reproductive output.

FROM NEURAL CIRCUITS TO SOCIAL CIRCUITS: COMING TO TERMS WITH THE MANY LAYERS OF COOPERATIVE BEHAVIOR

Cooperative behavior offers a number of challenges to researchers trying to understand why and how it evolved. One challenge is that it is difficult to untangle cooperative behaviors from their larger context within sociality. In the literature on cooperation there's a tendency to treat it as an isolated phenomenon. Yet cooperation is intimately and intricately tied to a larger suite of prosocial affiliative and helping behaviors. The physiological and neurological mechanisms that underlie cooperation may also underlie other prosocial behaviors.

Along these lines, psychologist Shelley Taylor notes in *The Tending Instinct* that altruistic behavior is so important to survival that nature has "backstopped" the behavior by wiring it into several different neurocircuitries. Writing about humans, Taylor observes that altruism may be so fundamental "that it has taken root in neurocircuitries for aggression, caregiving, and dominance, and from our capacity for bonding." Oxytocin, vasopressin, endogenous opioid peptides, and growth hormones make up what she calls the affiliative neurocircuitry, "an intricate pattern of co-occurring and interacting pathways that influence many aspects of social behavior." In mammals, oxytocin functions in milk letdown and labor, maternal care, mother-infant bonding, pair bonding, sexual behavior, and the capacity to form social attachments. Oxytocin facilitates prosocial behavior by lowering the natural resistance that animals have toward being in proximity to others. Although oxytocin may have evolved to promote the mother-infant bond, it seems to function more broadly now in cultivating cooperative behavior by fostering social closeness and trust. What Taylor's work suggests is that we cannot consider cooperation in isolation from other prosocial behaviors.

Cooperation is often paired with "affiliative" behavior, which strengthens social bonds or allows animals to exist in peaceful proximity. Grooming, for example, is affiliative in the sense that at least two animals have to be in close proximity, and it is cooperative in the sense of there being a reciprocal exchange of favors. Affiliative behavior creates the conditions in which cooperation can flourish.

As with sociality in general, there are various levels at which cooperation can take place: dyadic relationships between related or unrelated individuals, large-group networks (schooling fish), families (a prairie dog colony), small tightly knit groups (a wolf pack), and so forth. Cooperation can take place within an organism (cells cooperate within organs), within a society, and within an ecosystem. Cooperation can be either simultaneous (let's all work together now, as in a group hunt) or sequential (you groom me now and I'll groom you later). It can occur over a period of a few seconds, or be stretched out over years. Understanding cooperation, then, requires attention to these various levels of interaction.

It's often very difficult to move from the observation of a behavior to a secure conclusion that this behavior is indeed an instance of cooperation. The ethological literature is full of observations of animals appearing to help each other or work toward a common goal. Wolves, for example, are seen running together in pursuit of an elk in what appears to be a beautifully choreographed strategic dance. One wolf weaves left, another weaves right, one stays dead center. Together they take down an elk far too large for any single wolf to take individually. After the kill they take turns on the carcass, typically eating according to rank, higher-ranking individuals having priority. Did these wolves cooperate in the hunt? It is very hard to know. Unfortunately, observation does not lead seamlessly to explanation. Some scientists believe that these wolves are working together with a unified goal in mind, while others argue that the wolves may be acting quite independently of each other. They coordinate their actions because they know that the probability of taking down an elk alone is small. There's also a more minimalist explanation: the wolves just happen to coordinate their actions because each one is hungry and in search of food. This chance interaction is nothing more than individuals pursuing their own goals.

We have noted that not all ethologists and biologists agree that cooperation among animals is really cooperation. The fact that groups of chimpanzees hunt together and appear to coordinate their positions in the trees to most effectively trap and capture prey doesn't necessarily lead to the conclusion that they're cooperating. Like the wolves, they may be acting independently and simultaneously, without any cognitive decision to work together, however unlikely this may seem. Yet what this boils down to is really a matter of defining our terms. Cooperation skeptics, as we might call them, don't want to label animals as cooperative because this seems to accord animals too much cognitive power, too much intention, too much of all sorts of stuff that animals clearly (in the minds of these skeptics) do not have. Yet perhaps cooperative behavior is being drawn too narrowly. After all, the fact that a group of humans is working together to accomplish a goal doesn't require a discreet cognitive choice of the part of these collaborators. A great deal of the "why" of social interaction falls under the radar, as it were. We don't plan to cooperate or do an explicit calculation of its benefits; we just do it. Nor are we aware of the continual assessment of facial expressions and tones of voice that we unconsciously use to maintain a cooperative mood. The same is true for animals.

ULTIMATE AND PROXIMATE EXPLANATIONS FOR COOPERATIVE BEHAVIOR: THE THEN AND THE NOW

In the above example of cooperative hunting in wolves there are two levels of explanation an ethologist might pursue. She might wonder about what the wolves are doing now. She might look for the *proximate* explanation: What immediate goal is each animal pursuing, and what internal mechanisms are guiding its behavior? What are the cognitive and affective underpinnings of the behavior? What is the trigger stimulus? For example, a proximate trigger might be a pursuit invitation signal given by the elk, such as stotting, a bouncing yo-yo-like gait, which seems to say to a predator "catch me if you can." On the other hand, our ethologist might be interested in an *ultimate* explanation, seeking to understand why cooperative hunting has evolved and how it contributed to the reproductive fitness of an individual wolf.

A good deal of the literature on cooperative behavior has focused on the second type of explanation, attempting to understand how cooperation might have evolved "back then" and what would have made it a successful strategy for an individual or a group. The dominant ultimate theories about cooperation are kin selection, mutualism, and reciprocal altruism.

An additional ultimate explanation for social cooperation that we should mention briefly here is what evolutionary biologists call group selection. In group selection the focus of selection is the entire group, which flourishes, survives, or perishes as a whole. It is easy to see why group selection is so appealing in discussions of such phenomena as cooperation. It seems intuitive that a more cooperative pack of wolves would do better than a less cooperative group, in the sense that the group would survive and there would be more reproduction. Cooperative predation and defense of food yields more for the group, and the absence of food can lead to the group's dissolution. Despite its intuitive appeal, however, group selection remains controversial because of the strong influence of Darwinian theory, in which the focus of selection is individual fitness, rather than the survival of the group in which an individual lives. We believe, along with other biologists such as David Sloan Wilson and Edward O. Wilson, that group selection may regain status as a useful paradigm for understanding the evolution of cooperation and other prosocial behaviors.

While evolutionary explanations are very helpful in understanding why we see certain patterns of behavior in extant animals, evolutionary accounts can also give a false sense of what we truly know and understand. Many behavior patterns have complex origins and can persist in a behavioral repertoire for any number of biological and other reasons (psychological or sociological, for example). Of course, it's likely that a variety of evolutionary mechanisms favored the evolution of various sorts of social interactions, and right now we're learning which ones apply and when. What we're giving you is the "state of the art" with the caveat that these theoretical explanations are likely to evolve over time, as biologists collect more data and gain deeper understanding into social behavior. With that caution in mind, let us review the main ultimate theoretical explanations for the evolution of cooperation.

THE EVOLUTION OF COOPERATION

Darwin puzzled over certain behaviors that didn't seem to fit his proposed theory of evolution through natural selection. He hypothesized that individual fitness was the key to survival, but as he looked around he saw that various types of animals, including humans, formed closely knit social groups. Individuals worked together toward common goals and even engaged in behavior that seemed to reduce personal fitness in exchange for collective success. Although Darwin offered several reasonable explanations for this seemingly anomalous behavior, it wasn't until the 1960s that solid theoretical work on cooperation began and experimental evidence was gathered that could help elucidate the evolutionary mechanisms at play.

There are now several robust explanations for how cooperation might have arisen. It might be that individuals cooperate with or help relatives because conferring benefits on relatives is one mechanism for (partial) reproductive success: this is called *kin selection*. It could also be that cooperative behavior evolved because it conferred individual benefits on cooperators themselves. The theories of *mutualism* and *reciprocal altruism* both try to explain the direct benefits of cooperation. Each of these three explanations is probably correct; cooperative behavior may have evolved in a number of overlapping and interconnected ways. Let's look at each in turn.

IF YOU SMELL LIKE ME YOU MUST BE MY RELATIVE: HAMILTON'S KIN SELECTION

The renaissance of interest in the evolution of altruism and cooperation that occurred in the 1960s can be largely credited to the seminal work of William D. Hamilton, who died prematurely in 2000 of malaria that he contracted while in Congo. At the time he contracted malaria, Hamilton was researching the hypothesis that the initial spread of the human immunodeficiency virus (HIV) had been from nonhuman primates to humans.

Hamilton's early papers, published in 1964, began a revolution in evolutionary studies of animal behavior because he offered the first rigorous explanation of altruism. Hamilton, like Darwin, was especially interested in the evolution of altruism, in which one individual suffers a loss of reproductive fitness when it provides aid to another. Hamilton stressed the importance of *kin selection* in evolution, a process by which blood relatives who share a certain percentage of genes, genes that are identical by descent from a common ancestor, show a preference for one another rather than for unrelated individuals. Cooperation and altruism are expected to predominate when relatives interact with one another. For example, an individual may prefer to provide food to brothers or sisters instead of nonrelatives because his siblings share a higher percentage of genes than do unrelated individuals. Or, relatives might be more likely than nonrelatives to warn one another when a predator is nearby or to provide care to offspring to whom they're not genetically related.

There are many studies showing that kin selection is a strong force in the evolution of cooperation and altruism. One of the most famous examples of kin-selected altruism is Cornell University biologist Paul Sherman's study of alarm-call behavior in Belding's ground squirrels. Alarm calling carries a cost because it increases the likelihood of an individual being spotted by a predator. It turns out that males, who don't nest near genetic relatives, sound the alarm less often than females, who do live in proximity to genetic relatives.

Stuart West, Ido Pen, and Ashleigh Griffin carried out another interesting study, which provides strong support for Hamilton's theory. They showed that for fifteen species of birds and three species of mammals, helpers consistently discriminate between nuclear family members and more distantly related kin, and that there is stronger discrimination in species for which the benefits of helping are greater. The more one has to gain by helping relatives, the better he or she will be at making discriminations among other individuals.

If you find yourself wondering how animals know who exactly are kin and how closely related to them other individuals are, don't underestimate their powers of detection. For example, many animals recognize kin through an incredibly nuanced sense of smell, and kin smell different from nonkin. Siblings who are reared in the same nest acquire one another's odor, and this facilitates kin recognition. Olfactory recognition is mediated genetically by what's called the major histocompatibility complex (MHC).

However, individuals can be fooled into thinking that another individual is a relative, and this is why the system isn't perfect, even in nature. In the biblical story of Jacob and Esau, Jacob steals the blessing that is meant for his older brother. Their blind father Isaac is fooled into thinking that Jacob is really Esau, because Jacob donned Esau's clothes and Isaac "smelled the smell of his garments, and blessed him and said, 'See, the smell of my son.'" Smell can cause mistaken identity in animals, too. Vanderbilt University researcher Richard Porter and his colleagues Michael Wyrick and Jan Pankey discovered that if an infant spiny mouse from litter A is covered with the odor of a spiny mouse from litter B, then individuals in litter B will accept the spiny mouse from A as if she's from B. Similarity in odor trumps true genetic relatedness. Still, tricks such as these don't happen enough in nature to override the "armpit effect"—if you smell like me you must be kin.

Kin selection is well established empirically and is also supported by a large body of theoretical (especially mathematical) modeling; it can make successful predictions about behavior. Yet although Hamilton's work provides a good explanation for cooperative and altruistic behaviors among related individuals, it doesn't account for cooperative relationships among unrelated individuals. And there appear to be many instances in nature of unrelated conspecifics and even different species cooperating with each other. Two hypotheses have been put forward to explain why animals might help or cooperate with nonkin: reciprocity (also called *reciprocal altruism*) and mutualism (also called *by-product mutualism*). Both explanations are theoretically sound and both appear to explain at least some aspects of cooperative behavior (the two are not mutually exclusive). Still, there is significant disagreement about whether certain specific examples of cooperation should be labeled mutualism or reciprocity. We'll describe each hypothesis and say a bit about the disagreement, because it is relevant to our interest in moral behavior.

Before we continue, let us be clear that kin-selected altruism is not necessarily moral behavior. The sentinel behavior of the Belding's ground squirrel is not moral, nor is the self-sacrificing behavior of the unicellular social amoeba *Dictyostelium purpureum*, which directs altruism toward kin. This is not to say that altruistic acts are never moral, and we'll talk later in the chapter about when they might be, but most of the time they aren't. For now, we are simply noting possible mechanisms for the evolution of cooperation.

MUTUALISM: YOU SNOOZE, YOU ALL LOSE

Mutualism is a form of cooperation in which two or more individuals work together on a task that can't be accomplished singly. All those involved receive immediate benefit(s). It's a "you snooze, you all lose" situation; individuals rely on one another to the extent that they all lose without cooperation. Lee Dugatkin, a biologist at the University of Louisville and one of the leading researchers in animal cooperation, considers mutualism the simplest and probably most common type of cooperation: no kinship is necessary, and complex cognitive mechanisms (e.g., being able to keep score) are not required as they are for reciprocity.

Mutualism seems to be functioning in many species that engage in group hunting. Wild African dogs, lions, and wolves hunt in coordinated groups and when they hunt in a group they do better than single individuals in capturing and eating the large ungulates on which they feed. Even if a single animal is motivated solely by selfish interests, mutualism can evolve. Other cooperative behaviors in which mutualism seems to function include the joint defense of territory or resources, the formation of alliances, grooming, huddling, and vigilance against predators. Robert Sussman, Paul Garber, and James Cheverud argue that when primates cooperate (and we can also include cooperation in other species), benefits and costs don't need to be equal among all participants for mutualism to evolve. As long as the costs of cooperation are low, cooperative behavior could still evolve even if the rewards are also relatively small.

Mutualism can also occur between animals of different species. A recent example comes from the work of Redouan Bshary and his colleagues on cooperation between groupers and moray eels. The two species hunt together, and their combined hunting strategies are highly effective. The groupers swim to where the eels are resting in a crevice and begin to rapidly shake their head. The eels then emerge and they swim off together to hunt. Bshary's team was able to show that the cooperation was not random, because the groupers actually signal to the eels to initiate a joint hunt. The researchers were also able to show that both eels and morays benefited from the exchange.

Mutualism, then, consists of animals working together toward a common goal but doing what they would have done individually. There appears to be no conscious "choice" to cooperate, nor a complex calculation of whether cooperation is "worth it." Where there does seem to be some choice or calculation about the possibility of future benefits, the explanatory mechanism is not mutualism, but reciprocity.

The theory of *reciprocal altruism* was first proposed in 1971 by evolutionary biologist Robert Trivers. Trivers hypothesized that an individual might cooperate with or help another individual if the favor is later paid back. I'll scratch your back now, even though it is costly to me, with the expectation that you will scratch my back later. There is an important temporal element in the exchange when the payback is not immediate, as it is in mutualism. This can be risky, of course, since the recipient could decide to "cheat" and fail to repay the favor. Reciprocity thus requires a mechanism to deal with cheaters: there must be some way to detect cheaters and punish them appropriately. In long-lived social groups with little turnover in membership, reciprocal altruism based on future payback and detection of cheaters is theoretically possible. Nonetheless, true examples of reciprocal altruism in animal societies are thought to be rare.

Reciprocal altruism is difficult to test in animals because it's hard to know whether animals in the wild are genetically related, so it is often near impossible to rule out kin selection as the basis of cooperation. It is also extremely difficult to discern where exactly benefits and costs are cashing out, and especially to calculate how a particular behavior translates into future reproductive success or failure.

Despite these difficulties, a number of examples of reciprocity have been catalogued, notably in primates. For example, various primates are known to trade grooming for grooming, a phenomenon called *allogrooming*. Grooming patterns among a group of primates are not random, but follow a kind of "you scratch my back and I'll scratch yours" logic: grooming is reciprocal. For example, UCLA anthropologist Joan Silk and her University of Pennsylvania colleagues Robert Seyfarth and Dorothy Cheney found that female savanna baboons spend the most time grooming females from whom they have received the most grooming. Cheney and Seyfarth also showed that vervet monkeys are more likely to help individuals who have groomed them in the past. Grooming may serve a number of different functions, from the very practical task of removing parasites to the less tangible benefits of touch and physical proximity,

FIGURE 3. Luxe forms a grooming chain with her daughters Bex and Naxos. Photograph by Anne Engh.

which reduce social tensions and create a sense of bonding. Grooming, then, is not only a form of trading favors, but also a kind of cooperation aphrodisiac: it puts animals in a cooperative and affiliative mood, and may thus foster sociality. Researchers Dominic Johnson, Pavel Stopka, and David McDonald have suggested that an early step in the evolution of sociality may lie in "the shared pursuit of parasites rather than the more dramatic examples of cooperative hunting, or shared parental care of offspring."

Studies done on impala by University of California at Davis researchers Ben and Lynette Hart offer interesting nonprimate examples of reciprocity in grooming. The Harts have found that there is a high degree of reciprocity in allogrooming among impala. The benefits of grooming include the removal of ticks, the presence of which can lead to sickness, but grooming also has its costs, namely the reduction of vigilance and loss of electrolytes due to increased salivation. It turns out that regardless of gender or age, each individual in a grooming pair receives about the same number of cleanings as he or she initiates. Young fawns demonstrate reciprocal allogrooming, so it's reasonable to conclude that there has been strong selection for this behavioral trait.

Sometimes the "benefits" being traded are of different kinds, other than grooming itself. Louise Barrett and her colleagues discovered in the De Hoop Nature Reserve that adult female chacma baboons without infants would trade grooming for the privilege of holding another female's infant. Based on their findings, they suggested that grooming was a marketable commodity within some chacma communities. Similarly, Kathy Slater from Liverpool's John Moores University and her colleagues Colleen Schaffner and Filippo Aureli found that spider monkeys exchanged affiliative behavior, particularly hugging, for the privilege of handling an infant.

Grooming is one of the best-studied behavioral examples of reciprocity, but there are other documented cases of reciprocity in animals. One of the most vivid comes from biologist Gerry Wilkinson's research on vampire bats. Vampire bats leave their roost each night to forage for blood, which they generally drink from livestock. Some bats inevitably fail to eat, a dangerous failure because the bats need to eat almost nightly in order to survive. Those who are successful will share with those who are not. Wilkinson's research showed that bats share most readily with those who shared with them in the past.

Lee Dugatkin's research on the phenomenon of "predator inspection" in fish, especially guppies and sticklebacks, has been offered as yet another example of reciprocity. "Predator inspection," a term made up by the behavioral ecologist Tony Pitcher and his colleagues, refers to the way in which fish move slowly and with jerky movements away from their school and towards a potential predator, presumably to test whether the predator is hungry. Dugatkin studied pairs of inspectors to see if they behave cooperatively in a manner consistent with what is known as the tit-for-tat strategy, and the data suggest that this is the case. Inspectors begin inspecting at about the same time. They are, in biological lingo, "nice," and they stop inspecting if their partner stops, at which time they retaliate. Furthermore, it appears that inspectors remember who's who and prefer to hang out with cooperators rather than with cheaters, but they don't hold a grudge against the latter fish. It's interesting and perhaps surprising (but we think admirable) that Dugatkin's article was published in the prestigious journal *Nature* with the title "Trust in Fish."

As we see it, reciprocity can be a particular form of cooperation, but does not always involve cooperation. For example, if Marc brings Jessica a salt bagel laden with vegan cream cheese on Monday (which he does), and Jessica brings Marc one on Wednesday as a form of payback (perhaps anticipating that Marc will again bring her a bagel the following week), this is reciprocity, but not necessarily cooperation. They never worked together to achieve a common goal.

At the same time, not all complex cooperation is reciprocal. Researchers Robert Heinsohn and Craig Packer studied territorial conflicts in a group of female African lions. In order to simulate intrusion by members of another group of lions, Heinsohn and Packer played a tape of aggressive vocalizations that they had previously recorded. They found that some females, the "lead" females, were very active in approaching these "intruders," whereas others lagged behind. The lead females recognized that other females were laggards, but they did not punish them for not pulling their weight. Heinsohn and Packer concluded that the complex cooperative strategies shown in African lions are not necessarily maintained by reciprocity.

In addition to showing that reciprocity is not always a part of active cooperation, Heinsohn and Packer's paper raises a question about punishment in animal societies. We mentioned in chapter 1 that punishment, including such actions as third-party sanctions against those who break social rules, may be an important clue to moral behavior in animals. In the theoretical literature on cooperation, punishment for noncooperation is one of the key mechanisms thought to impact behavior. Timothy Clutton-Brock and Geoff Parker, for example, modeled punishing strategies in animals using evolutionary game theory and showed that punishment could be an important behavioral strategy for maintaining dominance relationships, discouraging cheats, disciplining offspring or potential sexual partners, or maintaining cooperative behavior. Unfortunately, the literature on punishment in animals is quite thin, and we have little more than intriguing questions. Future research on this important topic will require a mixture of ethology, cognitive psychology, evolutionary biology, and even philosophy. Indeed, this interdisciplinary approach is what is needed to delve deeper into cooperation and its many behavioral relatives.

There's one final piece to add to the reciprocity story. Two experiments have suggested that some animals may display what biologists call "generalized reciprocity," which up until recently was thought to be a uniquely human behavior. Felix Warneken, Brian Hare, and their colleagues at the Max Planck Institute for Evolutionary Anthropology observed that chimpanzees spontaneously and repeatedly helped humans who were trying to retrieve a stick from inside the chimpanzee's enclosure regardless of whether there was a reward. The chimpanzees also helped another chimpanzee gain access to a room in which there was food by removing a chain on the door. Recall also the research of Claudia Rutte and Michael Taborsky suggesting that rats display generalized reciprocity, providing help to an unfamiliar and unrelated individual based on their own previous experience of having been helped by an unfamiliar rat. In both cases, scientists believe that generalized reciprocity has occurred. Although suggestive, these studies have important limitations: both were conducted with small groups of animals in a captive setting, and the animals were asked to perform a behavior unlikely to occur in the wild. Much more research, especially studies on animals in their natural setting, is needed before generalized reciprocity can be confirmed.

Before moving on, let's quickly take stock of where we are. We've looked broadly at what kinds of cooperation animals exhibit, noted that there's a lot of cooperation in a wide variety of species, and shown that cooperation refers to a large collection of behavior patterns (kin-selected altruism, mutualism, reciprocal altruism, and generalized reciprocity). We've also reviewed several different evolutionary mechanisms by which cooperation could evolve. However, we still need to know more about what animals are experiencing when they reciprocate or cooperate, and what cognitive and emotional processes are at work. Only then can we leap into the moral arena and explore whether and when cooperative behaviors constitute moral behaviors.

MORAL EMOTIONS: THE AFFECTIVE FOUNDATIONS OF COOPERATION

Let's consider now what affective and cognitive skills are associated with cooperation. Note that we're moving from ultimate to proximate questions, from the then to the now. The transition is not totally seamless, for proximate and ultimate questions are hard to disentangle completely, but we're interested now in exploring what we know about the physiological mechanisms underlying cooperative behaviors. As in the other clusters, we're combining what we know about animals with what we know about humans and looking for cross-pollination of ideas.

Biologist Richard Schuster at the University of Haifa in Israel found that certain animals display a "bias to cooperate"—they seem to cooperate much more readily and more often than theoretical models predict they should. Schuster argues that we cannot look only at immediate outcomes of cooperation, because long-term consequences may be what drive the evolution of a behavior. One particular cooperative behavior may not have any fitness benefit for an animal, but cooperating in general has benefits. Schuster uses the example of lions. Hunting alone, a lion may get more food than hunting in a group, if he or she is able to acquire it. However, when these same animals also work together in defense of territory and cubs, cooperation suddenly has a great deal more importance. Because cooperation may not yield immediate material gain and has, instead, a long-term adaptive significance, there must be affective states that motivate and reward animals for cooperating.

What are some of the psychological mechanisms that might underlie or reward cooperative behavior? Since it's long been assumed that animals do not have emotions—or at least complex, interesting emotions—

there's little research that directly speaks to the emotional mechanisms on which cooperation and altruism rest. But we do know that in animals, emotions shape behavior in ways that enhance fitness. We also have a large body of research on the role of emotions in cooperation among humans. Given continuities in the architecture of human and animal psychology, comparative work may offer insights for further investigation of animal behavior.

Perhaps the most basic emotion that motivates cooperative behavior is affiliation—a sense of liking and a feeling of social closeness. Affiliation arises not only out of familial relationships, but also out of long-term pair-bonding (love) and friendship. Animals who live in close proximity may do more than simply tolerate the presence of others, they may actually enjoy social contact. The reverse is also true. Ample research attests to the fact that social animals who are isolated, either in zoos or research settings, become depressed and stressed.

We also know that endogenous opioid peptides (EOPs) foster affiliative and cooperative behaviors. Low levels of EOPs lead animals to seek social contact, and positive contact in turn leads to the release of EOPs. Neurobiologist Jaak Panksepp has suggested that EOPs may be responsible for a kind of social addiction: when animals are isolated, EOP levels are low, and animals crave social contact. When animals engage socially, they get a "hit" of EOPs, which creates a feeling of euphoria.

Might it feel good to cooperate? Yes, there are data that show that it does. We often are filled with warm feelings when we cooperate. Recent neural imaging research on humans by James Rilling and his colleagues shows that mutual cooperation is associated with activation of the brain's reward processing centers, the dopamine system. Our brain releases dopamine when we cooperate, giving us instant pleasurable feedback and reinforcing the behavior. This is significant research for it posits that being nice is rewarding in social interactions and might in itself be a stimulus fostering cooperation and fairness.

A research team led by University of Zurich economist Michael Kosfeld and his colleagues hypothesized that oxytocin might play a role in human approach behavior, specifically in our willingness to trust. Kosfeld's team created a "trust spray." They found that injecting volunteers with a nasal spray containing oxytocin made them more trusting.

An increased level of oxytocin led to an increase in trusting behavior. At least one company is already marketing Liquid Trust. Although not essential, trust is certainly important to human cooperation; it is the cornerstone of friendship, love, family, and trade. It is likely that trust plays a similarly foundational role in cooperation among animals. In a comprehensive review of the evolution of cooperation in animals published in 1981, Robert Axelrod and William Hamilton hypothesized that animals are more likely to cooperate with those whom they trust, and the complex cooperative relationships found in animal societies likely rest upon a foundation of stable, enduring relationships.

Other emotions are also likely important for greasing the wheels of cooperation in animal societies. Some that seem to play an important role in cooperation are anger (elicited by actual or perceived harm, such as failure to reciprocate), gratitude for a benefit received, forgiveness, empathy, spite, and envy. Evidence that animals get angry is indisputable. Less research has been done on more complex social and moral emotions such as gratitude and shame, but there's good reason to expect that animals with moral intelligence are capable of a broad range of emotional states that nurture and service the whole suite of moral behaviors.

COGNITIVE FOUNDATIONS OF COOPERATION: WHAT KINDS OF BRAINS DO COOPERATORS NEED?

Cooperation, like other facets of animal behavior, emerges from the interplay between external events happening to and around an individual—his or her animate and inanimate environment—and the individual's internal psychological and physiological milieu. We've looked at one major component of this "internal milieu," namely the emotional signals and experiences that shape behavioral responses. Let's turn now to the other principle component, the cognitive mechanisms that underlie these behaviors. Of course, cognitive and emotional mechanisms are intertwined and it is impossible to completely separate the two. But for the sake of discussion, we distinguish certain cognitive skills that facilitate cooperative behavior, particularly in its more complex manifestations.

In particular, animals need brains that can tie together past and present and make good guesses about the future. They also need brains that can make reasonably accurate assessments about the intentions and emotional states of other animals, both friends and strangers. They must be able to anticipate the behavior of a social partner, which involves "mentalizing"—attributing independent mental states to others, seeing them as distinct social actors with thoughts and emotions different from the animal's own. Animals must also possess considerable behavioral flexibility, such as being able to choose or suppress a certain course of action, based on an assessment of its likely outcomes.

Interestingly, the mental capacities that allow for reciprocity and complex cooperation are largely the same as those that function during competition, particularly in complex forms of deception and manipulation. Jean Decety and his colleagues have argued that *social cognition*, the mechanisms involved in understanding and interacting with others, evolved out of the dynamic interplay between opposing social forces of cooperation on the one hand, which can enhance fitness through greater security and better access to resources, and competition on the other, which can enhance fitness because it can afford an individual a selective advantage in reproduction or eating.

It's quite beyond controversy that animals have the mental skills to cooperate. This much is obvious, given the ubiquity of cooperative behaviors in the animal kingdom. And certainly many forms of cooperation require only relatively simple cognitive capacities. Kin-selected altruism and mutualism can both be found in a wide range of animals, including fish, birds, and insects. There is, though, an area of significant controversy, and this revolves around whether animals have the cognitive capacities necessary for the more complex forms of cooperation, such as reciprocal altruism and generalized reciprocity. This, of course, is of particular relevance to us, since we want to consider these more complex cooperative behaviors part of *wild justice*.

Biologists tend to view reciprocal altruism as the cognitive pinnacle of cooperative behavior, and some have concluded that only humans are capable of such flexible, nuanced, and complex behavior. Harvard researchers Jeffrey Stevens and Marc Hauser, for example, have taken this line and have argued outright that animals lack the cognitive mechanisms necessary for reciprocal exchange, and that they're not really nice to one another. These mechanisms, according to Stevens and Hauser, include numerical quantification, learning, memory, the ability to estimate time, and the use of reputation as a mechanism for assessing potential partners. Stevens and Hauser are certainly correct to note that reciprocity involves complex cognitive skills, and they may even be right that animals do not possess these skills in as sophisticated a form as humans. Nevertheless, the jury is still out on the question of whether animals have the cognitive skills necessary for complex forms of reciprocity, and just what these cognitive capacities may be. After all, the scientific understanding of social cognition in animals is still very young, and almost all of the limited comparative research on reciprocal altruism in animals focuses on primates.

GOING BEYOND THE PRIMATE PARADIGM: AVOIDING COGNITIVE SPECIESISM

Scientists and lay people alike tend to jump to conclusions about animal cognition, based on what is known about primates. For example, if primates, especially great apes, don't possess a particular cognitive skill, scientists often assume it doesn't exist elsewhere among animals, because they are all "less cognitively evolved" than primates. But this is not rigorous science. Rather, it's cognitive speciesism, the denial of certain cognitive skills to entire groups of animals based on little more than inaccurate stereotyping. Along these lines, Christine Drea and Laurence Frank note that researchers often are hesitant to see complex forms of cooperation in animals other than primates, and they raise the following important point concerning comparative research, one that speaks to cognitive speciesism: "Either the cognitive implications attributed to primates evincing cooperation should be extended to other animals, so that species solving similar problems are recognized as possessing at least comparable skills, or we should consider the possibility that the solution of such tasks reveals little about higher-order cognitive function." We favor their first suggestion.

It is important to remember that the way in which common behavior patterns are expressed may be unique in different species. For example, canids and felids tend to use visual signals and rapid and subtle exchanges of information to resolve social conflicts, whereas conflict resolution in rodents tends to be simpler and involve olfactory cues. Just as we have noted that there are unique forms of reciprocity in humans, different species of animals will display reciprocity and cooperation differently. So research into these behavior patterns needs to take place along a broad taxonomic spread, using methods and theoretical models appropriate to the species under study. We need, in particular, to move beyond the primate paradigm and keep our minds open to the possibility that nonprimate species may have evolved cooperative behaviors every bit as complex and adaptively fine tuned as chimpanzees and humans. For example, a study by Annemieke Cools and her colleagues Alain van Hout and Mark Nelissen called into question the assumption that reconciliation and third-party affiliations were unique to primates. Their study showed that social mechanisms used for peacemaking in dogs rivaled those of primates. Recall also Drea and Frank's work on cooperation in spotted hyenas that we mentioned in chapter 1. These hyenas engaged in behavior that just wasn't considered possible in a nonprimate. Indeed, Drea and Frank have had a difficult time getting their work published in peer-reviewed journals because readers were convinced that hyenas simply can't behave in such ways. Bernd Heinrich has also experienced trouble publishing his important data on corvid cognition because of the narrow-mindedness of reviewers. Likewise, Rutte and Taborsky's study on reciprocating rats challenged a stereotype long held sacred by scientists, but it did find a publisher and a good one at that. The bottom line here is that we must avoid cognitive speciesism, making decisions based on an outdated, linear evolutionary scale on which there are "lower" and "higher" animals.

Comparative research between and among species is also going to be crucial to understanding the breadth and subtlety of the cognitive mechanisms involved in cooperative behavior. Brian Hare's work, for example, shows that we cannot make generalizations even about primate cooperation because of the lack of consistent primate patterns in social living. Hare compared bonobos and chimpanzees engaged in the same cooperative task. When given a plate of food, a pair of bonobos will react by playing with each other and rubbing genitals (a behavioral reaction to social stress); they tend to share the fruit. A pair of chimpanzees will usually not share, and will avoid contact with each other. In a collaborative task where a team of two had to pull ropes to retrieve a plate of fruit (similar to Drea and Frank's hyena task), both the chimpanzee and the bonobo team worked together *if* the food was cut into small pieces that could be shared. But when the fruit was presented in large pieces, the chimpanzees cooperated less often, and when they did work together, one animal would try to monopolize the reward. This reminds us, again, that behavior will be species-specific.

COOPERATION AS MORAL BEHAVIOR: IS BEHAVIORAL FLEXIBILITY ENOUGH?

We've defined morality as a suite of other-regarding behaviors that cultivate and regulate complex interactions within social groups. So when is cooperation really "moral" behavior? As with moral behavior in general, we suggest that there will be a broad spectrum of cooperative and altruistic behavior, ranging from the very simple to the extraordinarily complex. We need to return to our threshold requirements to determine which altruistic and cooperative behavior patterns fall within the moral suite. We also need to use threshold requirements to distinguish which species fall within our narrower group of moral animals, in terms of their cooperative behaviors.

We need to be careful about language and remember that altruism has a specific meaning within biology and isn't synonymous with morality. Some researchers have claimed that slime mold behaves altruistically. For example, Richard Hudson and his colleagues published a report in the *American Naturalist* on "Altruism, Cheating, and Anti-Cheating Adaptations in Cellular Slime Molds." This is technically correct; in the cellular slime mold, some individual cells "sacrifice" themselves to become part of the stalk of the slime mold, which must die in order to support the live cells. It is worth quoting Hudson and his colleagues to make the point that researchers working on "lower" organisms do use the language of altruism and cooperation. They write "cellular slime molds (CSMs) possess a remarkable life cycle that encompasses an extreme act of altruism."

Despite the use of moral jargon, we wouldn't want to use the label "moral." If indeed slime molds were behaving altruistically, we wouldn't want to call it *moral* altruism, because slime molds don't meet our threshold requirements. Presumably, slime molds do not have rich emotional lives, nor do they have cognitive skills such as reading intentions or making predictions about the future. At the other end of the spectrum, our moral animals engage in social relations that are nuanced and complex and that require emotional and cognitive complexity, as well as behavioral flexibility. In our moral animals, we expect to see that altruism and cooperation are the bedrock of their sociality. We also expect to find a high level of emotional and cognitive complexity and flexibility. The more complexity and behavioral plasticity involved in a cooperative or altruistic behavior, the more "advanced" it is; the more likely it is morality.

Our proposal is that the moral animals are those capable of complex cooperative behaviors, and not just the simpler forms of kin-selected altruism and mutualism. This is consistent with the threshold requirements we laid out in chapter 1: a level of complexity in social organization, including established norms of behavior to which attach strong emotional and cognitive cues about right and wrong; a certain level of neural complexity that serves as a foundation for moral emotions and for decision making based on perceptions about the past and the future; relatively advanced cognitive capacities (a good memory, for example); and a high level of behavioral flexibility. Candidates would include animals such as bonobos, chimpanzees, elephants, wolves, hyenas, dolphins, whales, and rats.

The debate about whether and which animals have reciprocal altruism is certainly important. But reciprocal altruism is only one type of cooperative behavior, and other forms of cooperation may involve equally complex, though different, mental and emotional capacities. So even if we conclude that only chimpanzees are capable of reciprocal altruism, this isn't the end of the story as far as wild justice is concerned. Other cooperative and altruistic behaviors may also be just as subtle and nuanced.

To have a complete picture of animal morality we need to move on through the next two chapters, because the clusters of moral behaviors in our framework are closely interconnected: moral animals are capable of the whole range of behaviors and it's important to look at the whole picture. We'll see in the next chapter that at least some altruistic behaviors emerge out of an animal's capacity for empathy. For example, elephants display kindness toward each other, such as the outpouring of help for an injured or sick herdmate. And we'll see in the following chapter that complex forms of cooperation such as reciprocal altruism are closely tied to a capacity for fairness.

4 EMPATHY
MICE IN THE SINK

CeAnn Lambert, director of the Indiana Coyote Rescue Center, witnessed a small act of heroism in a sink in her garage. Two baby mice had become trapped in the sink overnight, unable to scramble up the slick sides. They were exhausted and frightened. Lambert filled a small lid with water and placed it in the sink. One of the mice hopped over and drank, but the other seemed too exhausted to move and remained crouched in the same spot. The stronger mouse found a piece of food. He picked it up and carried it to the other. As the weaker mouse tried to nibble on the food, the stronger mouse moved the morsel closer and closer to the water until the weaker mouse could drink. Lambert created a ramp with a piece of wood and the revived mice were soon able to scramble out of the sink.

Like the story of the two dogs and the meaty bone, one of them "nice" to the other, this story gets us thinking. What happened in the sink? Did one mouse actually understand that the other mouse was in trouble and find a way to help? Did the tiny creature display a kind of empathy? It's tempting to write off stories of this sort as an overexcited imagination reading far too much intention and emotion into the behavior of animals. Yet it is also possible to read too little into the animals we watch. Perhaps mice have the capacity to feel sorry for another mouse in distress, and to offer help. We'll never know about the mice in the sink, and the level of sympathy, intentionality, and understanding suggested by Lambert's account might seem unlikely in a rodent. Still, research may surprise us.

Indeed, in addition to innumerable stories, there's mounting scientific evidence that animals, even rodents, have the capacity to feel empathy. In June of 2006, researchers reported in the journal *Science* the first unequivocal evidence for empathy between adult nonprimate mammals. Dale Langford of McGill University and her colleagues demonstrated that mice suffer distress when they watch a cagemate experience pain. Langford and her team injected one or both members of a pair of adult mice with acetic acid, which causes a severely painful burning sensation. The researchers discovered that mice who watched their cagemates in pain were more sensitive to pain themselves. A mouse injected with acid writhed more violently if his or her partner had also been injected and was writhing in pain. Not only did the mice who watched cagemates in distress become more sensitive to the same painful stimuli, they became generally more sensitive to pain, showing a heightened reaction, for example, to heat under their paws. The researchers speculated that mice probably used visual cues to generate the empathic response, which is interesting since mice normally rely most heavily on olfactory communication. Although Langford's research falls far short of corroborating the mice in the sink tale, it nevertheless challenges some basic assumptions about mice and morality.

Other researchers were quick to note the importance of these unanticipated findings. Frans de Waal said of Langford's research, "This is a highly significant finding and should open the eyes of people who think empathy is limited to our species." These data confirm that empathy is an ancient capacity, probably present in all mammals. Jaak Panksepp remarked, "If it turns out that the 'empathetic' effect in mice is mediated by the same brain mechanisms as human empathy, then the evidence would be truly compelling that Langford's model actually reflects evolutionary continuity in a pro-social mechanism among many different mammalian species."

Many of us believe in the essential goodness of humans, and find confirmation of this belief in the everyday and seemingly random acts of kindness perpetrated not only by our family and friends, but also by strangers. We like to believe that our tendencies toward empathy and kindness are stronger than our tendencies toward cruelty and meanness. Can we also entertain the notion that in animal societies this same tendency toward kindness and compassion might be present? There is solid evidence that many animals have a capacity for empathy, and that empathy is a basic regulator of social life for at least some species of animal. In addition to countless anecdotal accounts, there are empirical data from ethology and neuroscience that confirm what many of us already know: animals are empathic creatures, with a large capacity for fellow feeling and behavior that reflects strong social attachments that endure over time.

De Waal and Panksepp, longtime students of animal behavior, seem unsurprised that mice show empathy. What neither says outright, but is implicit, is a more startling possibility: if animals share with humans the capacity for empathy, they have in place the cornerstone of what in human society we know as morality. For among humans, the capacity to understand what another feels allows us to be compassionate, to avoid causing pain or suffering, and to act with an intention to improve the welfare of those around us.

WHAT IS EMPATHY? THE LEXICON OF FEELING

Empathy is the ability to perceive and feel the emotion of another. As such, our empathy cluster includes sympathy, compassion, caring, helping, grieving, and consoling. The word *empathy* (from the Greek *em* "put into" + *pathos* "feeling") was coined in the early twentieth century, to translate the German word *Einfühlung*, which means literally "feeling into." The term *empathy* first appeared in the context of art, and referred to the capacity of a person to mentally identify with, and thus fully comprehend, an object of contemplation, a painting perhaps, or a piece of music. Yet the word quickly made its way into the lexicon of psychology, where it became (and remains) a concept of considerable interest as well as disagreement. In this context, the word refers to the capacity to read and understand the emotions of others and respond in a sensitive and helpful manner. The appearance of *empathy* in the literature on animals is a bit more difficult to trace; the word seems to have emerged sometime in the 1960s or '70s, but only recently has it become a subject of focused research and discussion.

The term *empathy* can be confusing because its meaning often shifts from one discipline to another, and little effort is made to carefully define how the word is being used. Philosophers, for example, frequently use words like *empathy* and *altruism* differently than evolutionary biologists. Philosophers have mainly written about sympathy, while biologists have written about empathy (though Darwin himself used the term *sympathy*). There is also confusion about the distinction between sympathy and empathy, particularly when working across disciplines. We define sympathy as a "feeling for," and empathy as a "feeling with." When you feel sympathy for another person, you don't necessarily share their emotion; when you empathize, you do.

Ultimately, it will be useful to have clear terminology that carries the same meaning in biology, ethology, human psychology, neuroscience, and other related fields; this will aid our attempts to understand evolutionary continuity of emotions and social behaviors. Most of the work in ethology has focused on empathy, and very little has been written about sympathy in animals. Our central concern will thus be with exploring empathy. We hope that future research will help elucidate the distinctions between empathy and sympathy in animals, and will explore both of these and other related phenomena.

EMPATHY — FROM SIMPLE TO COMPLEX

The most careful and successful attempt so far at defining and clarifying the meaning of empathy in relation to animals can be found in the work of Stephanie Preston and Frans de Waal. They define empathic behaviors as those in which one individual comes to perceive or understand the emotional state of another individual, through a "shared-state mechanism." *Shared state* means that empathy is by definition an intersubjective experience. The essence of empathy is emotional linkage. As Preston and de Waal explain, "The emotional state of one individual has the potential to elicit a similar state in nearby individuals. This emotional linkage has been present in primitive forms through much of the evolutionary history of chordates in the form of alarm and vicarious arousal. This basic linkage was then augmented by enhanced cognitive and emotional abilities through evolution and extended ontogeny (development of the individual), allowing individuals to experience empathy in the absence of releasing stimuli, towards more distant individuals, and without being overwhelmed by personal distress."

Empathy, as Preston and de Waal suggest, isn't a single behavior, but a whole class of behavior patterns that exists across species and shows up with varying degrees of complexity. It occurs in nested levels, with the inner core a necessary foundation for the other layers. The inner core consists of relatively simple forms of empathy such as body mimicry and emotional contagion, which are largely automatic physiological responses. The next layer consists of somewhat more complex behaviors such as emotional empathy and targeted helping. More complex still is cognitive empathy, or the ability to feel another individual's emotion and understand the reasons for it. Finally, and most complex, is the capacity for attribution, in which an individual can fully adopt the other's perspective, using the imagination.

Evolution of course doesn't toss out one adaptation and replace it with another. Rather, during the course of evolution, modifications are made to existing structures and capacities and these changes tend to reflect the social and environmental conditions to which individuals are exposed. More complex forms of empathy such as cognitive empathy evolved from emotional contagion, which, in turn, probably evolved from emotional linkage of individuals, especially emotional linkage between mother and infant. All empathy behaviors, from simple to complex, likely share many proximate mechanisms.

The notion of empathy as nested levels mirrors a more general aspect of mammalian evolution. Paul MacLean hypothesized that the mammalian brain is actually three brains in one (a triune brain, he called it), each successive stratum having been formed on top of the layer beneath it. Each layer of the brain has its own function, though all three are interconnected and interacting. The primitive brain, which MacLean called the reptilian or R-complex, is charged with the task of physical survival, controlling breathing and heartbeat and generating the fight-or-flight response. The limbic system, or paleomammalian brain, controls emotion. And the neocortex, the outer and most recent part of the brain, allows for higher cognitive functions such as language and abstract thought. The three layers function independently, but are also interconnected and interdependent in complex ways that are only partially understood. So, while emotional contagion may be in some respects simpler and more primitive, and may arise from older parts of the brain, it is most likely inextricably connected with more complex, more cognitive forms of empathy. Cognitively advanced forms of empathy are probably influenced in some measure by the more primitive and automatic impulses.

Some scientists deny that animals have empathy. But they usually come to this conclusion because they have narrowly cast empathy to mean the ability to take the perspective of another. The capacity for imaginative attribution may, indeed, be found only in humans. Yet this is but one small piece of a much larger group of behaviors, many of which are certainly present in a broad range of mammals. And as de Waal and Preston argue, it is premature to pronounce animals lacking in cognitively complex forms of empathy because we still know too little about empathy in animals to make such a claim. Cognitive empathy may be found, for example, in the hominoid primates and perhaps elephants, social carnivores, and cetaceans.

WHY EMPATHY IS ADAPTIVE

Emotional contagion is an emotional state in one individual that results directly from perceiving the emotional state of another. When someone yells "Fire!" in a crowded movie house, the panic is contagious. No one may actually see or smell fire, but the fear and panic are palpable and move people to do something. And during times of social unrest, mobs are dangerous precisely because energy and anger can flow through a crowd so quickly that violence on a large scale can erupt in response to a seemingly small or isolated provocation. In humans, emotional suggestion is a powerful shaper of social behavior. We're exquisitely tuned in to the body language, facial expressions, and tone of voice of those around us, and will unconsciously mimic and synchronize these outward expressions of emotion. When someone across the table yawns, we will likely yawn in turn and not even notice that we've done so. If the person we're talking to holds his arms crossed in front of his chest, we will likely cross ours as well.

Other social animals are similarly linked emotionally and take behavioral cues from the emotional state of others in their social network. Watch at the dog park as one dog spots a new arrival and begins to bark. The other dogs will begin a frenetic barking, and only after the chorus of barks is well under way will the dogs look around to see what they're barking at.

Or watch the birds in your backyard. If one bird startles and flies off, others will follow, not waiting around to assess whether the threat is real. They have been infected by emotional contagion. In a long-term research project that Marc did with some of his students on patterns of antipredatory scanning by western evening grosbeaks, a highly social finch, they found that birds in a circle showed more coordination in scanning than did birds who were feeding in a line. The birds in a line, who could only see their nearest neighbor, not only were less coordinated when scanning, but also were more nervous, changing their body and head positions significantly more than grosbeaks in a circle, where it was possible for each grosbeak to see every other grosbeak. Marc wondered whether the birds in line were more fearful because they didn't know what their flockmates were doing. Emotional contagion would have been impossible for individual grosbeaks in the linear array except with their nearest neighbors.

Animals living in social groups can benefit from being sensitive to the emotional states of other group members. Emotional contagion might, for example, facilitate defensive action in light of threat. If one prairie dog gives an alarm call, all members of the group will respond immediately with evasive action. The same holds true for birds: if you startle one bird away from the feeder, most if not all of the birds will disperse. And not only will all the sparrows fly off, but likely so will the robins, grackles, and finches, suggesting that emotional contagion may function between species. This behavior spreads out the costs of vigilance, allowing individuals more time to forage, mate, or care for young.

We most often think of fear and panic as being contagious, as with the flock of geese that burst into flight when one gets scared. But joy, excitement, curiosity, and intense interest can spread quickly as well. Social play is often so contagious that it appears epidemic. For example, when a dog sees other dogs playing he or she often spontaneously jumps into the fray, and dogs also emit what's been called a play pant, or "laugh" that spreads the play mood among other dogs who hear this vocalization. In a study of emotional contagion in orangutans, Marina Davila Ross and her colleagues studied play behavior in twenty-five orangutans between two and twelve years of age. They discovered that when one of the orangutans displayed an open, gaping mouth, the orangutan's equivalent of human laughter, its playmate would often involuntarily display the same expression less than half a second later. Roger Highfield, writing in the UK *Telegraph*, noted that facial mimicry, a building block of emotional contagion, predates humans by many millions of years, since we share a common ancestor with orangutans some twelve to sixteen million years ago. Along these lines, Matthew Gervais and David Sloan Wilson, working at the state University of New York in Binghamton, have suggested that human laughter might also be important in "playful emotional contagion."

Emotional linkage between individuals can lead to forms of empathic response in which the observer perceives the emotional state of another and "feels sorry for" this emotional state. Empathy may just remain a feeling state (I feel distressed to see you so distressed), but it also may motivate some action, such as trying to alleviate the source of distress or offering comfort. Empathy may thus be an important component of certain altruistic and cooperative behaviors. In particular, it may facilitate complex cooperative interactions such as reciprocal altruism. It may also function in the development of trust, since trust involves being able to assess the intentions and emotions of interaction partners. Of course, the ability to read and understand intentions also facilitates manipulation and deception, and the capacity to imagine how one's own behavior affects others can lead to extreme forms of cruelty.

THE COSTS OF EMPATHY

Evolution is a balancing act between costs and benefits, cashed out ultimately as an individual's reproductive success. Empathy seems at first glance like a win-win behavior, particularly if the empathic response involves only an affective reaction and no particular helping response.

Yet empathy can be costly in various ways. These ways have not yet been explored in much depth in the empathy literature, but might run along the following lines.

Researchers Jean Decety and Philip Jackson call attention to what we might term the cost of an expanded self. A self that is linked to others also shares in the emotional experiences of others. When we see someone in distress, we too feel distress and perhaps anxiety. When we see someone experiencing fear, we too feel fear. And distress, anxiety, and fear are not "free"; they demand cognitive and metabolic attention, and can divert attention and energy from important tasks. Fear, panic, and distress cause the brain to release cortisol, the "stress hormone." The release of cortisol in the body sets off a cascade of physiological effects: blood pressure goes up, digestion stops, pulse quickens. Too much cortisol in the body can lead to impaired cognitive function, lowered immunity, and other costly changes. This is why misplaced empathy or too much empathy might be maladaptive. Too much of a good thing can be bad.

Empathy may be costly not only for the empathizer, but also for the individual who is the object of empathic response. Humans and animals alike may benefit from being able to hide emotions such as our excitement at finding a huge cache of great food or our fear during a struggle for dominance. The better those around us are at reading our facial expressions, tone of voice, body language, and olfactory messages, the less successful we will be at masking our intentions and feelings. The capacity for empathy creates in a society of animals a level of transparency and intersubjectivity that makes honest communication the norm and may explain why deception is considered more cognitively demanding than honesty.

THE FACIAL ECOLOGY OF EMPATHY: WOLVES, DOGS, AND FOXES

Research conducted by ethologist Michael W. Fox on facial expressions in wolves, coyotes, and red foxes sheds some light on species differences in emotional connectedness, emotional contagion, and empathy. Facial expressions are likely a good indicator of social (and, we would argue, moral) complexity: the more facial displays, the greater and more nuanced the social information that can be communicated. Wolves are highly social carnivores, much more so than either coyotes or red foxes. Wolves also have much more complex facial displays then either coyotes or red foxes. And wolves, according to our moral taxonomy, would likely have more highly developed moral capacities and nuanced empathy, perhaps displayed as communicating and responding to more subtle variations of social rules, than coyotes or red foxes.

Our discussion of what has been called "the facial ecology of empathy" also relates to synchronicity in human behavior that's related to how we perceive another person's emotional state. Empathy is not mediated by cognitive or conscious evaluative mental processes, but is "pure" interaction; we read people's expressions and through this have a fairly accurate understanding of the emotional state they are experiencing.

SO WHAT DO WE REALLY KNOW?

The science of empathy in animals is really quite young, and ethologists are in the early stages of exploring empathic capacities in animals. Indeed, some scientists remain skeptical about empathy in animals. Nevertheless, there is some highly suggestive narrative and empirical evidence for empathy in elephants, several cetacean species (especially bottlenose dolphins and toothed whales), rats and mice, social carnivores, and primates. This suggests that empathy is distributed across many species. We have no doubt that continued research in this area will expose a richness and depth of empathy in a wide range of social mammals.

As with the other clusters, the evidence for empathy comes from the convergence of many different streams of research, especially ethology, psychology, and neuroscience. Some of the most intriguing pieces of the empathy puzzle come from research on humans. New ideas about animals may emerge as we study human behavior, particularly if we remain attuned to evolutionary continuity. Sometimes the connection to empathy in animals is quite inadvertent. For example, psychologist Carolyn Zahn-Waxler was studying the responses of young children to the distress of a family member. So she went into the homes of a number of families and observed how children reacted to parental distress. The behavior of the household pet turned out to be just as interesting as the behavior of the child. When a family member feigned sadness or distress—when he or she pretended to cry or choke—the household dogs would often show more concern than the children, hovering nearby or nudging their owners, or gently resting their head on the distressed person's lap.

We can also hope to learn more about human behavior by studying empathy in animals. As we'll discuss below, the discovery of mirror neurons in monkeys is leading researchers to a deeper understanding of empathic behavior in humans (and has also uncovered a new window into understanding autism-spectrum disorders). There's great interest in the neuroscience of empathy, which will help elucidate the cognitive and affective mechanisms at play.

EARLY INTIMATIONS: A BIT OF HISTORY

Charles Darwin suggested that human morality is an extension of social instincts, and that human morality is continuous with similar social behavior in other animals. He paid special attention to the capacity for sympathy, which he believed was evidenced in a large numbers of animals. Darwin tells a number of stories, including these about birds: "Capt. Stansbury found on a salt lake in Utah an old and completely blind pelican, which was very fat, and must have been well fed for a long time by his companions. Mr Blyth, as he informs me, saw Indian crows feeding two or three of their companions which were blind; and I have heard of an analogous case with the domestic cock." Darwin calls sympathy an essential part of, and indeed the foundation stone for, other social instincts. He concludes, "Any animal whatever, endowed with well-marked social instincts, the parental and filial affections being here included, would inevitably acquire a moral sense of conscience, as soon as its intellectual powers had become as well-developed, or nearly as well-developed, as in man."

Darwin emphasized that the differences between humans and other animals—in all realms, including the moral sentiments—were of degree, not of kind. Darwin, it turns out, was quite right about the importance of the sentiments, about the role of sympathy, and about the evolutionary continuity between humans and other social animals. Yet his ideas lay mostly dormant for more than a century.

WITNESSING EFFECTS: MORE ON RODENT EMPATHY

In 1959, long before Langford's discovery of empathic mice, Russell Church, a researcher at Brown University, published an essay in the *Journal of Comparative and Physiological Psychology*: "Emotional Reactions of Rats to the Pain of Others." Church had set up an experimental test in which rats were trained to press a lever in order to get a food reward. He then set up, in a neighboring cage, a torture chamber of sorts: the bottom of the cage was an electric grid on which the rats' delicate pink paws were placed. When a rat in the first cage pressed the food lever, a surge of electricity would run through the grid in the adjoining cage, giving the neighboring rat an electric shock. Church found that rats would not push the food lever if they could see that a fellow rat would receive a shock. Although Church himself did not explain the reaction as empathy, this seems in retrospect to be the most parsimonious explanation.

Another early study in 1962 by George Rice and Priscilla Gainer titled "'Altruism' in the Albino Rat" showed that rats would help other rats in distress. One rat was suspended in air by a harness and a neighboring rat could press a level to lower the suspended rat. The suspended animal would typically squeak and wriggle in distress. The rats were apparently made uncomfortable by signs of distress in a fellow rat, and would act to alleviate the distress by pressing the lever. Empathy likely motivated the "altruistic" response. Although little research has since focused on empathy in rodents, Langford's surprising discoveries about mouse empathy will probably revitalize interest in these animals.

One related area of research worth mentioning is the phenomenon known as "witnessing effects." Jonathan Balcolmbe, Neal Barnard, and Chad Sandusky summarized numerous studies indicating that mice and rats show a marked stress response to being in the same room as another rat subjected to decapitation. Rats show increased heart rate and blood pressure (both stress responses) when watching other rats being decapitated, and when a paper towel with dried blood from a decapitated rat is placed atop their cage. Witnessing effects have also been documented in mice, monkeys, and of course humans. As Balcombe noted in a comment on Langford's mouse-empathy research, witnessing effects clearly arise out of the capacity for empathy, and add additional support to the data on empathy in rats and mice.

Studies of empathy in animals often are horribly cruel, and it is deeply ironic that we inflict pain on other animals to test them for empathy when good evolutionary biology—evolutionary continuity—tells us that they have it. It is also ironic that the animals most frequently used in research—mice and rats—presumably because they have less going on "up there" or "in there" than primates, turn out to have quite a bit more going on inside than researchers have assumed. Although noninvasive ethological fieldwork can certainly provide data on animal empathy, it is likely that invasive laboratory research will continue. The findings on rat and mouse empathy will suggest ways in which we can make these studies—not just empathy research, but all research in which rats and mice suffer, particularly when they suffer in the presence of others—more humane and less stressful. After all, general stress levels experienced by laboratory animals compromise the reliability of the data, a point made by University of Arizona physiologist Ann Baldwin and Marc.

EMPATHY IN PRIMATES

Let's now consider "higher" animals, who supposedly do have a lot going on "up there" and "in there." People who went to the 2007 "Mind of the Chimpanzee" conference held in Chicago, Illinois, were abuzz over a chimpanzee named Knuckles. Knuckles is the only known captive chimpanzee with cerebral palsy, which leaves him handicapped both physically and mentally and unable to function as a normal member of his chimpanzee group. What's surprising about Knuckles is not just that he himself manages to survive with a debilitating disease, but that the other chimpanzees in his group treat him differently. The community apparently understands that Knuckles is different, and adjust their behavior accordingly. Although a young male would normally be subjected to intimidating displays of aggression by older males, Knuckles is rarely subjected to such treatment. Even the alpha male is tolerant of Knuckles and grooms him gently. Knuckle's friends empathize with him and as a result treat him kindly.

Knuckles's story is but one instance of empathy in chimpanzees. In an interview, anthropologist Barbara J. King told the story of Tina and Tarzan.

> A chimpanzee female named Tina was killed by a bite to the neck by a leopard. She'd been living in a community of chimpanzees for quite a long time. The group didn't just pull at her body or tug at it or ignore it. Rather, the dominant male of the group sat with her body for five hours. He kept away all the other infants and protected the body from any harm with one exception. He let through the younger brother of Tina, a five-year-old called Tarzan. That's the only youngster who was allowed to come forward. And the youngster sat at his sister's side and pulled on her hand and touched her body. I think this was not just a random occurrence. The dominant male was able to recognize the close emotional bond between Tina and Tarzan, and he acted empathically.

Anyone who has worked with chimpanzees knows that they are empathic beings, and that the stories of Knuckles and of Tina and Tarzan aren't all that surprising. Indeed, the most robust research on empathy in animals comes from the primate literature. It may be that primates, of all the social mammals, have the most highly developed empathic abilities. Or it might simply be that the sheer abundance of primate research yields a corresponding abundance of data on empathy, and the more carefully we look at other species, the more we'll find. At any rate, the research on primates is revealing, and begins to uncover many of the nuances of empathic behavior in animals.

Primate research carried out in the 1960s was suggestive, though in that era few scientists would have been willing to label any nonhuman behavior as truly empathic. A classic study published by Stanley Wechkin, Jules Masserman, and William Terris in 1964 showed that a hungry rhesus monkey would not take food if doing so subjected another monkey to an electric shock. The monkeys refused to pull a chain that delivered them food if doing so gave a painful shock to a companion. One monkey refused to pull the food chain for a full twelve days, starving itself seemingly to avoid causing pain to another.

Around the same time, University of Wisconsin psychologist Harry Harlow was setting forth on his famous wire-monkey experiments. Although Harlow was interested in humans, his controversial research on monkey love also revealed a great deal about the process of social attachment in primates, the very process that is thought to shape the neural connections that underlie empathic behavior. Working with infant rhesus monkeys who had been taken from their mothers, Harlow showed that the desire for affection was stronger than the desire for food. Given a choice between a cold wire monkey with food and a soft cloth monkey without food, the infants clung to the soft, foodless monkey. From other studies, Harlow concluded that baby monkeys raised without social contact with peers and without real mothers grow up to be socially incompetent. The development of social and moral intelligence is stunted when the appropriate developmental cues are not triggered. Harlow's work led to later studies on attachment and on the important connection between the early nurturing of infants and children and the development of empathy.

In another study conducted in 1977 by Hal Markowitz, diana monkeys were trained to insert a token into a slot to obtain food. A male was observed helping the oldest female, who had failed to learn the task. On three occasions he picked up the tokens she had dropped, put them into the machine, and allowed her to have the food. His behavior seemed to have no benefits for him; there did not seem to be a hidden agenda.

Although many of these early studies involved monkeys, there is now a large body of research that spans the range of primate species. And having the opportunity to compare empathic capacities in monkeys and apes reveals important differences, and confirms the hypothesis that empathy is seen in a broad range of behavioral tendencies and that species will vary, perhaps considerably, in how developed these capacities are. For example, Frans de Waal asserts that empathy is more cognitively complex and more highly developed in great apes (chimpanzees, bonobos, and humans) than in monkeys. He argues, as a case in point, that consolation behavior, in which one animal (a bystander) consoles another after a fight, is indicative of cognitive empathy. Consolation behavior has been demonstrated in great apes but not monkeys. Orlaith Fraser, Daniel Stahl, and Filippo Aureli showed that consolation in captive chimpanzees reduces stress in recipients of aggression (there are decreases in self-scratching and self-grooming, behavioral indicators of stress) and also that consolation may act as an alternative to reconciliation when reconciliation doesn't occur.

NEURAL UNDERPINNINGS OF EMPATHY: MIRRORS AND SPINDLES

Behavioral data leave no doubt that animals can display empathy. It's also useful to consider the neurobiological data on empathy. The discovery of mirror neurons in monkeys more than a decade ago led to a revolution in how scientists understand the connection between the brain and behavior, including empathic behavior. Mirror neurons fire when an animal performs an action and when the animal observes someone else performing the same action. Although research on mirror neurons is still relatively young, one hypothesis that has gained support is that mirror neurons may, among other things, play a functional role in empathy. They appear to be portals of empathy. Research on humans shows that mirror neurons or their functional equivalents are activated during the observation and imitation of social emotions, especially as these emotions are read through visual cues such as facial expression. Yawning when someone else yawns and wincing when we see someone hit their finger with the hammer are both activated by mirror neurons. In humans, mirror-neuron systems are thought to mimic actions and read intentions and emotions. We create a neural template in our own brain for someone else's action or for the emotion associated with the action.

Mirror neurons are a likely neural substrate for emotional contagion in a span of animal species, although which species actually have mirror neurons (or neurons that function similarly) remains largely undiscovered. Although research has connected mirror neurons to empathy in humans, much is still unknown. Whether there's also a connection between mirror neurons and empathy in animals, and in which species, remains uncertain. Still, there is every reason to believe that the brains of other animals work in a similar way. For example, Derek Lyons, Laurie Santos, and Frank Keil write about the sensitivity of nonhuman primates to the mental states of others, and the role of mirror neurons in enabling primates to infer the intentions of other agents.

Alongside mirror neurons, spindle cells appear to be crucial in empathy. Spindle cells, also called von Economo neurons, are a class of neural cells located in the prefrontal cortex (in humans, at least) that are thought to process social emotions and play an important role in social attachment. Spindle cells are not unique to humans, but have been considered unique to the hominoid line. They have been identified in chimpanzees, bonobos, orangutans, and gorillas, but not monkeys. This adds additional support to Preston and de Waal's thesis that empathy is less complex and nuanced in monkeys than in hominoids. Spindle cells may, like mirror neurons, be connected to autism disorders in humans. The location of spindle cells in the brains of autistic individuals appears to be abnormal, which may lead to impairment in social behavior, including decreased empathy.

In a surprise to scientists, spindle cells were recently identified in several species of toothed whale, including humpback whales, fin whales, killer whales, and sperm whales. Spindle cells appear to have existed in whales for at least twice as long as in humans, and whales have more of them. The discovery of spindle cells in whales is very exciting because it raises the possibility that empathy may be found in even a broader range of species than previously imagined.

EMPATHY BENEATH THE SURFACE: COMPASSIONATE CETACEANS

The discovery of spindle cells in whales adds to a growing literature on empathy in marine mammals, especially cetaceans. The cetacea include some ninety species of whale, dolphin, and porpoise. They are thought to be some of the most intelligent animals on earth, and also some of the most socially sensitive.

There are many anecdotal accounts from marine biologists of cetaceans displaying empathy. Mark Simmonds, an expert on toothed whales, tells of a pod of false killer whales that remained with an injured member of their group for three long days, in water so shallow that they were exposed to sunburn and risked becoming stranded. The pod stayed with the injured whale until he finally died. Simmonds also tells the story of two male orcas who appeared to be grieving over the death of their mother. After their mother died, the two males swam together, apart from other orcas in their pod, retracing their mother's movements on the last few days of her life. Naomi Rose, the cetacean researcher who witnessed this event, interpreted this as grieving. Orcas are also known to grieve for lost calves. There are many anecdotal accounts of dolphins showing empathy for other dolphins as well. Research on cetaceans also suggests a significant capacity for empathy, as noted by cetacean experts Kathleen Dudzinski and Toni Frohoff.

EMPATHY IN ELEPHANTS

Let's now return to land, where elephants have been observed to be extremely empathic. They are known for the tenderness they show to each other, and for their close-knit societies. There are countless anecdotal accounts of elephants showing empathy toward sick and dying animals, both kin and nonkin.

Joyce Poole, who has studied African elephants for decades, relates the story of an adolescent female who was suffering from a withered leg on which she could put no weight. When a young male from another group began attacking the injured female, a large adult female chased the attacking male, returned to the young female, and touched her crippled leg with her trunk. Poole concluded that the adult female was showing empathy.

Injured elephants also figure into other stories about empathy. Recall Babyl, the injured elephant whom Marc observed while out in the field with elephant expert Ian Douglas-Hamilton. Because of an injured rear leg, Babyl could only walk at a snail's pace, and for over a decade and a half, the other elephants in her group have waited for her and fed her. Unescorted, Babyl would easily have fallen prey to a lion. There's also the story of a forest elephant who had lost her trunk to a trapper's snare. The injured elephant learned how to drink and how to eat river reeds,

FIGURE 4. Grace of the Virtues family touches Eleanor of the First Ladies with her trunk and foot, before lifting Eleanor back onto her feet. Courtesy of Shivani Bhalla, from Iain Douglas-Hamilton, S. Bhalla, G. Wittemyer, and F. Vollrath, "Behavioural Reactions of Elephants towards a Dying and Deceased Matriarch," *Applied Animal Behaviour Science* 100 (2006): 87–102.

the only food she could manage without her trunk. Group members helped to keep their friend alive by altering their own feeding habits, and bringing her reeds. And now it has been reported that all this group eats is river reeds.

Ian Douglas-Hamilton, who has studied elephants for more than four decades, has observed numerous instances of empathy. In one, he describes how Grace, of the Virtues family, attended to Eleanor, matriarch of the First Ladies family. Eleanor was ailing, unable to stand steadily. When she fell, Grace gently touched Eleanor with her trunk and foot and then lifted her back to her feet. As Douglas-Hamilton writes in his field observation: "Grace tried to get Eleanor to walk by pushing her, but Eleanor fell again . . . Grace appeared to be very stressed, vocalizing, and continuing to nudge and push Eleanor with her tusks . . . Grace stayed with her for at least another hour as night fell." After Eleanor

FIGURE 5. Maui from the Hawaiian Islands family steps over and pulls at Eleanor's dead body. Courtesy of Shivani Bhalla, from Iain Douglas-Hamilton, S. Bhalla, G. Wittemyer, and F. Vollrath, "Behavioural Reactions of Elephants towards a Dying and Deceased Matriarch," *Applied Animal Behaviour Science* 100 (2006): 87–102.

died, a number of elephants visited the body, some touching and some just standing for a time near the dead matriarch. A female named Maui "extended her trunk, sniffed the body, touched it, and then tasted [Eleanor's] trunk. She hovered her right foot over the body, nudged the body, and then stepped over, pulling the body with her left foot and trunk, before standing over the body and rocking to and fro."

Elephants grieve openly for their dead. One story in the *Sunday Times* told of a baby elephant killed by a lioness. Over the course of the day, elephants from the herd gathered in a rough circle around the remains of the baby. Many of them touched the body with their trunks. Elephants also show a pronounced interest in corpses and bones, a behavior thought to exist only in elephants and humans. Karen McComb and her colleagues designed a study to investigate the concern that elephants show for the dead. They presented wild elephants with a collection of skulls and other objects. They found that the elephants spent more time smelling and feeling elephant skulls than they did the skulls of rhinoceroses or buffalo. However, when elephants and rhinos form close social relationships, elephants will mourn the loss of their rhino friends. Thus, in an incident that was reported in Zimbabwe in November 2007, after her black rhino companion was shot, dehorned, and buried by poachers, Mundebvu, an African elephant calf living in Zimbabwe, "dug down for about one metre to try to reach her former companion, constantly letting out screams and shrieks as the other two elephants supported her."

As Douglas-Hamilton says, with the understatement of a seasoned scientist, "the question of whether or not there might be compassion or suffering among surviving elephants who interact with ailing or dead ones remains so far unanswered." But, he continues, "observations suggest that this could be the case." It's reasonable to suppose that a capacity for empathy is associated with the expression of compassion for the ailing and grief for the dead.

SOCIAL BREAKDOWN IN ELEPHANT SOCIETIES: THE DEVASTATING COSTS OF EMOTIONAL TRAUMA

What shapes behavior—what allows empathy to flourish in a person and also in other animals—is social environment and early development, particularly maternal nurturing. Nature may plant the seeds of empathy—the neural circuitry that can develop from emotional linkage and attachment into empathy—but if the seeds are not nourished, development can go wrong.

An essay published in the journal *Nature* in 2005 by psychologist Gay Bradshaw and her colleagues on what they call "elephant breakdown" gives us a window into the connection between early experiences— especially maternal nurturing—and the development of empathy. The bonding process between mother and infant facilitates the development of neurophysiological structures that underlie normal social behaviors such as empathy. We know that in humans a disruption of this bonding process can result in reduced capacity for empathy and an increased propensity toward violence. Early trauma has permanent effects on the brain, and thus on behavior. Trauma such as separation of the infant from its mother, or abuse or neglect by the mother, can lead to a permanent impairment in empathic social interaction.

Bradshaw and her colleagues hypothesize that social disruption in animal societies, in this case elephants, has interfered with the normal development of young elephants, particularly by depriving them of appropriate maternal care and teaching. This early trauma can lead to empathic impairment in elephants, just as it can in humans. Elephants live in very tightly bonded matriarchal societies, with layers of extended family who participate in caring for and rearing young. In the early 1990s, there were an estimated ten million wild elephants. These populations have been decimated by poaching, culls, and habitat loss, and only about a half a million elephants now survive in the wild. The complex social structures of elephant society are collapsing under the weight of loss and fragmentation. Infants have been orphaned, often after witnessing their parents being brutally killed. Some of the remaining elephants, particularly young males, are displaying symptoms very much like human post traumatic stress disorder (PTSD): depression, abnormal startle response, unpredictable social behavior, and violent aggression.

Matriarchs are storehouses of social knowledge, and the loss of a matriarch can have wide-ranging impacts on elephant society. Bradshaw and UCLA neuroscientist Allan Schore report, "infants are largely reared by inexperienced, highly-stressed, single mothers without the socio-ecological knowledge, leadership, and support that a matriarch and allomothers provide." Most astonishing to researchers has been the killing of white and black rhinoceroses by young male elephants. These young males were cull orphans, or were born to mothers who had witnessed a cull, or were reared within socially disrupted herds. Not only has the normal process of maternal care been derailed, but also the larger social structure of elephant society has been disrupted.

When human societies disintegrate and the social fabric becomes damaged, people often lose their moral bearings. This may be equally true for animal societies held together by normative standards of behavior. This suggests, among other things, that in planning for conservation, we need to pay particular attention to conserving intact and functioning societies, not just saving individual animals.

EMPATHY AS A BUILDING BLOCK OF MORALITY

Let us take stock of what we know about empathy in animals. We know that the capacity for empathy has evolved in mammals who live in complex social groups, and that empathy helps foster and maintain social cohesion. There's evidence for empathy in primates, pachyderms, cetaceans, social carnivores, and rodents. The capacity for more nuanced and complex empathic behavior seems to be correlated with both social complexity and intelligence. Because empathy is grounded in the same neurological architecture as other prosocial behaviors such as trust, reciprocity, cooperation, and fairness, it seems likely that a whole suite of interlinked behaviors have co-evolved in social mammals. Empathy is possibly among the most basic of these prosocial behaviors, having evolved out one of nature's earliest experiments in social attachment: the mother-infant bond.

Here we arrive, then, at the startling implication of Langford's study of mouse empathy: humans may not be the only species with morality. Indeed, it is likely that morality has evolved in a number of species, in conjunction with and as an adjunct to sociality. The difference between the moral behavior of animals and that of humans is, as Charles Darwin suggested, a difference in degree, not kind.

FEELING BEYOND SPECIES LINES: IMPROBABLE FRIENDS

In Langford's study of empathy in mice, the mice's writhing reaction was especially pronounced if they had been cagemates, suggesting greater empathy for a familiar mouse than a stranger. Mice demonstrate what seems to be a larger truth of empathic concern: the empathic response is strongest in the center, and weakens as it radiates out. This same pattern of empathic preference for family and neighbor has been documented in many other species, including humans. We saw the same radiating pattern with cooperation and altruism.

But as the research we've explored in this chapter clearly demonstrates, animals can and do show empathy toward nonkin. They also, perhaps surprisingly, show empathy for members of other species. With respect to cross-species, animal-to-animal empathy, one of the most compelling stories comes from Frans de Waal. De Waal watched as Kuni, a captive female bonobo living at the Twycross Zoo in England, captured a starling and took the bird outside and placed it on its feet. When the bird did not move, Kuni tossed it in the air. When the starling did not fly, Kuni took it to the highest point in her enclosure, carefully unfolded its wings, and threw it in the air. The starling still did not fly, and Kuni then guarded and protected it from a curious juvenile. It seems clear that Kuni was taking the perspective of the bird.

Marc's late dog Jethro once brought home a tiny bunny, whose mother had likely been killed by a mountain lion near Marc's home. Jethro dropped the bunny at the front door and when Marc came to the door he looked up as if to say, "please help." Marc brought the bunny into his house, and put it into a cardboard box with water, carrots, and lettuce. For the next two weeks, Jethro was pinned to the side of the box, refusing to go out for walks and often missing meals. After Marc released the bunny, and for months afterward, Jethro would run to the spot and search around for it. Years later, Jethro saw a bird fly into the window of Marc's car and picked up the stunned ball of feathers and carried it over to Marc, once again seeming to ask for help. Marc placed the bird on the hood of the car, and a few moments later the bird flew off. Jethro watched attentively as it took flight.

The most striking example of cross-species empathy is the relationship between companion animals and their human guardians. There are also countless tales of animals helping humans, including stories of dolphins helping humans at sea. In New Zealand, a pod of dolphins was observed forming a protective circle around a group of swimmers to fend off an attack by a great white shark. Philosopher Thomas White tells of a dolphin named Tursi who changed her behavior when she discovered that a young boy was blind. Tursi herself was blind in one eye and White wonders if this had anything to do with the way in which she related to the boy. There's also a story of three lions in Ethiopia who rescued a twelve-year-old girl from a gang who had kidnapped her. Numerous stories about dogs helping humans emerged from the tragic events of 9/11 and the Asian Tsunami. And, of course, we have Binti Jua, who helped the young boy who fell into her enclosure at the Brookfield Zoo.

There's also a wonderful story of a young chimpanzee named Joni, raised by the Russian primatologist Nadia Ladygina-Kohts some eighty years ago. Joni often went onto the roof of Kohts's house, and calling for her, scolding her, or offering food didn't work to get her down. But crying did. To quote Kohts:

> If I pretend to be crying, close my eyes and weep, Joni immediately stops his plays or any other activities, quickly runs over to me, all excited and shagged, from the most remote places in the house, such as the roof or the ceiling of his cage, from where I could not drive him down despite my persistent calls and entreaties. He hastily runs around me, as if looking for the offender; looking at my face, he tenderly takes my chin in his palm, lightly touches my face with his finger, as though trying to understand what is happening, and turns around, clenching his toes into firm fists.

Empathy is foundational to morality, in humans and animals alike. We begin to see, with empathy, that our three clusters are closely interconnected. Threads from one reach into the others. For example, empathy weaves into cooperation and altruism, for, as you may have noticed, many of the acts of kindness and helping described in the chapter—acts motivated by empathy—are instances of altruism. Empathy is also connected to justice, and justice, in turn, to cooperation and altruism. Before we discuss these interconnections among the clusters, let us spend some time with animals who seem to have a sense of justice and fairness.

5 JUSTICE
HONOR AND FAIR PLAY AMONG BEASTS

"Fairness is only human, scientists find." This headline appeared in the *Los Angeles Times* as we were writing this very chapter. The study being discussed had recently been published in the prestigious journal *Science*, and had attracted a lot of attention. Keith Jensen and his colleagues at the Max Planck Institute devised what is called an "ultimatum game," a favorite tool of economists who study human decision marking. This sort of game involves two players, one of whom is given a small amount of money and asked to divide the money between themselves and their partner however they see fit. The partner knows how much money is being divided and if he or she receives too low an offer, one they think is unfair, the offer may be rejected and neither player will receive anything.

Jensen's study was unique because the players were chimpanzees and the currency was raisins. Jensen and his team found that the chimpanzees didn't play the game like humans typically do. In studies of human behavior, offers of less than 20 percent of the money are almost always rejected. In contrast, the chimpanzees accepted any offer from their partner and didn't get upset by low offers in which the chimpanzee offering the raisins kept most for himself.

In the summary of their research, the authors note, "These results support the hypothesis that other-regarding preferences and aversion to inequitable outcomes, which play key roles in human social organization, distinguish us from our closest living relatives." They concluded, in other words, that chimpanzees are not sensitive to fairness. Ironically, however, the behavior of these chimpanzees is considered more rational in pure economic/game-theory terms. In the *Los Angeles Times* article, lead author Keith Jensen said the chimps behaved more rationally than people because "it makes perfect economic sense to accept any nonzero offer and to offer the smallest amount possible while keeping the most for yourself."

JUSTICE ISN'T SOME BONE-IN-THE-SKY IDEAL

Jensen's research on resource sharing is fascinating, and offers a glimpse into what may turn out to be some very interesting differences in how human fairness behaviors differ from the fairness behaviors of other species. But the conclusion of the authors, that chimpanzees don't have a sense of fairness, doesn't follow from their work. The only conclusion we can safely draw from this specific research project involving an ultimatum game is that chimpanzees don't behave like humans, leaving wide open the question of whether chimpanzees have a sense of fairness.

Sarah Boysen, a primatologist at Ohio State University who was asked to respond to Jensen's research, drew a different conclusion than the researchers. She believes that chimpanzees have a strong sense of justice, though it is different from ours. Boysen notes, "Deviations from their code of conduct are dealt with swiftly and succinctly, and then everybody moves on." Research by Sarah Brosnan and Frans de Waal on inequity aversion in captive chimpanzees and capuchin monkeys offers support for Boysen's claim. So too does recent research by Friederike Range and her colleagues on inequity aversion in domestic dogs. We'll discuss this line of research below.

Jensen's experiment may open a window into the evolution of fairness and other-regarding behavior, but it should also serve as a cautionary tale. The few published studies that investigate fairness in nonhuman primates involve only a handful of animals, which limits our ability to gather information on individual variability. Furthermore, because these studies have been conducted over a short period of time, we're unable to get an appreciation for emerging patterns of behavior within a stable social group. That the animals are living in controlled captive conditions also may be a confounding factor as may be their having been asked to perform tasks that they don't typically perform in the wild. This isn't to say that the data are useless, but rather to stress that the negotiation of fairness among animals is a dynamic process that likely changes from one social situation to another.

JUSTICE IN ANIMALS OTHER THAN PRIMATES

Jensen and colleagues conclude that if the closest relative to humans, *Pan troglodytes*, doesn't have a sense of fairness, certainly no other animals will. Case closed. But the case is not closed, not by any means. Virtually all of the research on fairness in animals has been conducted on non-human primates. Yet there are other fascinating species such as wolves and coyotes and even domestic dogs from which we can gain insights into the behavior patterns that are used to negotiate fair deals. In fact, renowned philosopher Robert Solomon, in his book *A Passion for Justice*, asks us to consider pack-living wolves, exemplars of highly developed cooperative and coordinated behavior. Solomon writes:

> Some wolves are fair, a few are not. Some arrangements are fair (from the wolf's own perspective); some are not. Wolves have a keen sense of how things ought to be among them . . . justice is just this sense of what ought to be, not in some bone-in-the-sky ideal theoretical sense but in the tangible everyday situations in which the members of the pack find themselves. Wolves pay close attention to one another's needs and to the needs of the group in general. They follow a fairly strict meritocracy, balanced by considerations of need and respect for each other's 'possessions,' usually a piece of meat.

Solomon stresses the importance of learning more about wolves in his discussion of justice, emotions, and the origins of social contracts.

A major message of *Wild Justice* is that we need to look at animals other than nonhuman primates and study what they do when they interact with one another socially. In the spirit of open-minded science, let's give other animals a chance to show who they are, what they know, and what they feel. Closing the door, on ideological grounds, to the possibility that species other than primates have a sense of justice—that if nonprimates don't do something, then surely other animals don't either—means that we'll never come to appreciate the full array of behaviors shown throughout the animal kingdom.

We believe that a sense of fairness or justice may function in chimpanzee society, and in a broad range of other animal societies as well. While there is less research on justice than on cooperation and empathy, comparative data, especially on social play behavior, an area of research that hasn't been given much attention by primatologists, speaks to the question of the distribution of justice in nonhuman animals.

WILD JUSTICE

JUST: what is merited or deserved.

JUSTICE: the maintenance of what is just, especially by adjustment of conflicting claims or assignment of merited rewards or punishments. (Merriam-Webster)

Justice is a set of expectations about what one deserves and how one ought to be treated. Justice has been served when these expectations have been appropriately met. Our justice cluster comprises several behaviors related to fairness, including a desire for equity and a desire for and capacity to share reciprocally. The cluster also includes various behavioral reactions to injustice, including retribution, indignation, and forgiveness, as well as reactions to justice such as pleasure, gratitude, and trust.

The word justice doesn't have any special meaning in biology. One reason for the lack of a rigorous or even a semirigorous working definition is that very few studies have been conducted on justice in animals and there has been very little discussion of this phenomenon among evolutionary biologists and ethologists. As research accumulates, a vocabulary will inevitably evolve and it will be important to make choices about which terms most closely fit observed patterns of behavior.

We realize that discussing justice in animals might invite comments of the "Surely, you're joking" variety. But we're not. Despite splashy headlines to the contrary, researchers still don't know much about other animals' reactions to inequity and unfairness. But we feel confident that some animals do, indeed, have a sense of justice. Why do we make this claim while others have been hesitant to do so?

First of all, we argue from an evolutionary perspective stressing continuity. A sense of justice seems to be an innate and universal tendency in humans. Research from psychology, anthropology, and economics supports this conclusion. For example, research conducted by economists Ernst Fehr and Simon Gächter found that humans get inordinately upset about unfairness, and will even forgo immediate personal gain in order to punish a perceived injustice, as in the ultimatum game. Consider also that prelinguistic human babies show social intelligence that may provide the foundation for morality, and for a sense of justice, later in life. At six months of age, before they can sit or walk, human babies are able to assess another person's intentions, and these social evaluations are important in deciding who's a friend or who's a foe. In one study, an infant was shown a puppet show in which there was a nice or a nasty character who either helped or hindered a character trying to walk uphill. Afterwards, when the infants were encouraged to reach out and touch either the helper or the hinderer puppet, they chose the helper. Furthermore, the infants preferred the helper over a neutral character and the neutral character over the hinderer.

Kiley Hamlin, who along with her colleagues at Yale University conducted this study and published the results in *Nature*, noted, "We don't think that this says that babies have any morality, but it does seem an essential piece of morality." And, furthermore, "Our findings indicate that humans engage in social evaluation far earlier in development than previously thought, and support the view that the capacity to evaluate individuals on the basis of their social interactions is universal and unlearned." The authors also conclude that "social evaluation is a biological adaptation."

We agree with the general conclusions of Hamlin's study and offer that even in the absence of symbolic language, animals are able to make these sorts of social evaluations and that these assessments are foundational for moral behavior in animals other than humans. Indeed, recent research by Francys Subiaul of the George Washington University and his colleagues showed that captive chimpanzees are able to make judgments about the reputation of unfamiliar humans by observing their behavior—were they generous or stingy in giving food to other humans? The ability to make character judgments—generous or stingy—is just what we would expect to find in a species in which fairness and cooperation are important in interactions among group members.

The principle of parsimony suggests the following hypothesis: a sense of justice is a continuous and evolved trait. And, as such, it has roots or correlates in closely related species or in species with similar patterns of social organization. It is likely, of course, that a sense of justice is going to be species-specific and may vary depending on the unique and defining social characteristics of a given group of animals; evolutionary continuity does not equate to sameness.

Furthermore, fairness is not merely an overlay that masks competition and selfishness. Lee Dugatkin and Marc have shown, using game-theoretic models, that always acting fairly should be more common than never acting fairly and that continuing to be fair during social development can be an evolutionarily stable strategy (ESS). (An evolutionarily stable strategy is one that, if adopted by a population of individuals, is resistant to invasion by any alternative strategy.) So, like cooperation, fairness has played a significant role in the evolution of social behavior. It's not a dog-eat-dog world because really dogs don't eat other dogs.

Second, and even more central to our argument about justice in animals, are the data from animals themselves. Although little research has focused directly on the question of whether animals have a sense of justice, there are tantalizing clues from research on various other aspects of animal behavior. Our agenda here is to lay out for you these hints. We begin with social-play behavior, which offers the most compelling evidence for a sense of fairness in social mammals. In the context of play behavior, we can look at ways in which animals understand, communicate, and enforce a set of rules about fairness. We will then turn to the few studies of what researchers call "inequity aversion," because these studies have a direct bearing on our discussion of fairness and justice. Finally, we'll explore some of the behavioral reactions to fairness and injustice, including pleasure, indignation, trust, forgiveness, and retribution.

WHAT'S PLAY GOT TO DO WITH MORALITY?

Morality is rather like a game: there are agreed-upon rules that everyone must follow, and there are sanctions for breaking the rules. The rules are, in a sense, an imaginative construction. They are relative to the game at hand. In social groups, as in a game, the integrity of the collective depends upon individuals agreeing that certain rules will regulate their behavior. At any given moment individuals know their place or role, and that of other group members.

Social play, in turn, provides insights into morality. In particular it opens a window into behavior patterns included in our justice cluster. Social play is a voluntary activity that requires that participants understand and abide by the rules. It rests on foundations of fairness, cooperation, and trust, and it can break down when individuals cheat. During social play, individuals can learn a sense of what's right or wrong—what's acceptable to others—the result of which is the development and maintenance of a social group (a game) that operates efficiently. Thus, fairness and other forms of cooperation provide a foundation for social play. Animals have to continually negotiate agreements about their intentions to play so that cooperation and trust prevail, and they learn to take turns and set up "handicaps" that make play fair. They also learn to forgive.

Social play has unique rules of engagement about how hard one can bite, about mating being off limits, and about assertions of dominance being absent or kept to a minimum. Think about games such as tag or hide-and-seek or keep away. There are special rules that apply during these games, but not otherwise. Those joining the game must understand these rules (which are often implicit) and abide by them, lest they be labeled a cheater and expelled from the game. If players don't cooperate, play can easily escalate into fighting.

When animals play, they must *agree* to play. They must cooperate and behave fairly. Further, when fairness breaks down, play not only stops, it becomes impossible. *Unfair play* is an oxymoron, and this is why play is such a clear window into the moral lives of animals.

WHAT IS PLAY AND WHY DO IT?

Jethro bounds towards his dog friend, Zeke, stops immediately in front of him, crouches on his forelimbs, wags his tail, barks, and immediately lunges at him, bites his scruff and shakes his head rapidly from side to side, works his way around to his backside and mounts him, jumps off, does a rapid bow, lunges at his side and slams him with his hips, leaps up and bites his neck, and runs away. Zeke takes wild pursuit of Jethro and leaps on his back and bites his muzzle and then his scruff, and shakes his head rapidly from side to side. Suki bounds in and chases Jethro and Zeke, and they all wrestle with one another. They part for a few minutes, sniffing here and there, and then rest. Then Jethro walks slowly over to Zeke, extends his paw toward Zeke's head, and nips at his ears. Zeke gets up and jumps on Jethro's back, bites him, and grasps him around his waist. They then fall to the ground and mouth wrestle. Then they chase one another and roll over and play. Suki decides to jump in and the three of them frolic until they're exhausted. Never did their play escalate into aggression. This scene is taken from Marc's field notes.

Play behavior is a widespread phenomenon. When animals play, they use behavior patterns from a variety of other social contexts. For example, actions that are used in mating (mounting) are intermixed in highly variable kaleidoscopic sequences with behaviors that are used during fighting (vigorous biting), looking for prey (stalking), and avoiding being someone else's dinner (fleeing). Thus, social play can be confusing to the players themselves, and they need to know that play is the name of game as the encounter progresses.

According to University of Tennessee psychologist Gordon Burghardt, an expert on animal play, the evolutionary roots of play may go back over a billion years. There is evidence of play behavior in diverse phylogenetic groups, including placental mammals, birds, and even crustaceans. Of course not all animals play, but oddly enough, the animals of particular interest to us in this book, nonhuman primates, rodents, canids, felids, ungulates, pachyderms, and cetaceans, tend to be the most playful animals. Coincidence? Probably not.

Play is adaptive and serves important functions in diverse animals. In some, such as members of the dog family (dogs, coyotes, wolves, and foxes), play is important for the development of social skills and for the formation and maintenance of social bonds. During play, animals learn social norms and reciprocity. Play can also be practice for the "real thing," as when wolf cubs or male mountain sheep play fight. Play also provides physical exercise (aerobic and anaerobic, during which bones, tendons, joints, and muscles are used) and cognitive training (in the form of

FIGURE 6. Young coyote pups playing outside of their den in Yellowstone National Park, Wyoming. Courtesy of Thomas D. Mangelsen/Images of Nature.

"eye-paw" coordination). Marc and his colleagues Marek Spinka and Ruth Newberry, whose specialty is pig behavior, view play as training for the unexpected because play is a highly variable behavior and prepares individuals for rapidly changing and novel or surprising situations.

Neuroscientists and ethologists have argued that play creates a brain with more behavioral flexibility and greater learning capacity. During play there is a continual assessment of playmates' intentions and signals and respect for certain rules unique to play. When coyote cubs play, their behavior is variable and unpredictable. They jump from one kind of behavior to another, engaging patterns from various contexts ranging from reproduction, predation, and aggression, stimulating the brain and helping the brain draw connections. Thus, play is cognitively demanding and can be thought of as "brain food." It helps to rewire the brain, increasing the connections between neurons in the cerebral cortex. It hones cognitive skills including logical reasoning and behavioral flexibility and provides important nourishment for brain growth. Psychologist Stephen Siviy's research showed that bouts of play in rats increase

FIGURE 7. When dogs and other animals play they use behavior patterns from different contexts including fighting, hunting, and mating. Here, Sasha (left) rears while playing with her friend Woody, as she would if they were fighting. Sasha and Woody played vigorously and fairly over a period of five years, and only twice did their play spill over into mild fighting that lasted for about three seconds, after which they immediately continued to play. While playing they fine-tuned their behavior "on the run." From a video taken by Marc Bekoff.

the brain's level of c-FOS, a protein associated with the stimulation and growth of nerve cells.

University of Lethbridge psychologist Sergio Pellis, one of the leading researchers of animal play, even believes that larger brains are linked to greater levels of play. And researcher Kerrie Lewis, who has studied play in primates, has shown that primate species with greater levels of social play have larger neocortex size relative to less playful primates.

There is strong selection for playing fairly because most if not all individuals benefit from adopting this behavioral strategy. Group stability may be also be fostered through play. Numerous mechanisms including play invitation signals, variations in the sequencing of actions performed during play when compared to other contexts, self-handicapping, and role reversing have evolved to facilitate the initiation and maintenance of social play in many species of mammals.

Play is not only serious business, but also fun. Animals get deep joy and pleasure from playing alone or with friends. Rats emit a high-frequency chirp when they play wrestle and when they get tickled, a sound that some rat researchers describe as a laugh. Dogs laugh too. They make a kind of breathy, forced exhalation that is recognized by other dogs as an invitation to play. It feels good to laugh because it triggers the brain to release dopamine. The rhythm, dance, and spirit of animals at play are also incredibly contagious, and spread like an epidemic; just seeing animals playing can stimulate play in others.

FAIR PLAY: FINE TUNING ON THE RUN

The social dynamics of play require that players agree to play and not to eat one another or to fight or mate with one another. Play means play, and not fighting or mating. When there's a violation of these expectations, others react to this lack of fairness. For example, young coyotes and wolves react negatively to unfair play by ending the encounter or by generally avoiding those who ask them to play and then don't follow the rules. Coyotes and wolves who play unfairly find it difficult to get others to play with them after they've been labeled a cheater.

Domestic dogs also don't tolerate noncooperative cheaters, who may be avoided or chased from play groups. While studying dog play on a beach in San Diego, California, Alexandra Horowitz observed a dog she called Up-ears enter into a play group and interrupt the play of two other dogs, Blackie and Roxy. Up-ears was chased out of the group and when she returned, Blackie and Roxy stopped playing and looked off toward a distant sound. Roxy began moving in the direction of the sound and Up-ears ran off, following their line of sight. Roxy and Blackie immediately began playing once again.

Animals exhibit fairness during play, and they react negatively to unfair play behavior. In this context, fairness has to do with an individual's specific social expectations and not some universally defined standard of right and wrong. If you expect a friend to play with you and he acts in an aggressive manner, dominating or hitting rather than cooperating and playing, then you will feel you are being treated unfairly because of a lapse in social expectations. We have found, by studying the details and dynamics of social play behavior in animals, that animals exhibit a similar sense of fairness. For instance, one way we know that animals have social expectations is that they show surprise when things don't go "right" during play, and only further communication keeps play going. For example, during play when a dog becomes too assertive, too aggressive, or tries to mate, the other dog may cock her head from side to side and squint, as if she's wondering what went wrong. For a moment, the violation of trust stops play, and play only continues if the playmate "apologizes" by indicating through gestures such as a play bow his intention to keep playing.

We want to stress that social play is firmly based on a foundation of fairness. Play only occurs if, for the time they are playing, individuals have no other agenda but to play. They put aside or neutralize any inequalities in physical size and social rank. As we will see, large and small animals can play together, and high-ranking and low-ranking individuals can play together, but not if one of them takes advantage of its superior strength or status.

After all is said and done, it may turn out that play is a unique category of behavior in that asymmetries are tolerated more so than in other social contexts. Animals really work at reducing inequalities in size, strength, social status, and how wired each is to play. Play can't occur if the individuals choose not to engage in the activity, and the equality or fairness needed for play to continue makes it different from other forms of cooperative behavior (such as hunting and caregiving). Play is perhaps uniquely egalitarian. And if we define justice as a set of social rules and expectations that neutralize differences among individuals in an effort to maintain group harmony, then that's exactly what we find in animals when they play.

DON'T BOW IF YOU DON'T WANT TO PLAY

Let's look at the data for our claims about the connection between social play and morality. Most of the research on play and fairness has been

FIGURE 8. A play bow performed by the dog on the right. Marc measured the duration of individual bows and also their form on a grid system, where form was equal to the declination of the shoulders relative to standing height (a is the vertical displacement of the shoulders on a grid system). Bows are highly stereotyped and easily recognized actions that are used to signal "I want to play with you," "I'm sorry I bit you hard, let's keep playing," or "I'm going to bite you but it's only play." For details see this text and M. Bekoff, "Social Communication in Canids: Evidence for the Evolution of a Stereotyped Mammalian Display," *Science* 197 (1977): 1097–99; and M. Bekoff, "Play Signals as Punctuation: The Structure of Social Play in Canids," *Behaviour* 132 (1995): 419–29.

on domestic dogs and their wild relatives, coyotes and wolves. While we'll focus on the animals we know the best, there are also examples from other animals that support our views about the connection between social play and morality.

When dogs and their relatives play, they use actions that are also used in other contexts, such as dominance interactions, predatory behavior, antipredatory behavior, and mating. Because there's a chance that various behavior patterns that are performed during ongoing social play can be misinterpreted as being real aggression or mating, individuals have to tell others "I want to play," "This is still play no matter what I am going to do to you," or "This is still play regardless of what I just did to you."

Play frequently begins with a bow, and repeated bowing during play sequences ensures that play remains the name of the game. A dog asks another to play by crouching on her forelimbs, raising her hind end in the air, and often barking and wagging her tail as she bows. After each individual agrees to play and not to fight, prey on, or mate with the other, there are ongoing rapid and subtle exchanges of information so that their cooperative agreement can be fine tuned and negotiated on the run, so that the activity remains playful.

After many years of studying play in infant canids (domestic dogs, wolves, and coyotes, members of the dog family), Marc realized that the bow isn't used randomly, but rather with a purpose in mind. For example, biting accompanied by rapid side-to-side shaking of the head is performed during serious aggressive and predatory encounters, and can easily be misinterpreted if its meaning isn't modified by a bow. Not only are bows used right at the beginning of play to tell another dog "I want to play with you," but they're also used right before biting accompanied by rapid side-to-side head shaking as if to say "I'm going to bite you hard but it's still in play" and right after vigorous biting as if to say "I'm sorry I just bit you so hard but it was play." Bows reduce the likelihood of aggression.

Play signals are almost always used honestly. Cheaters who bow and then attack are unlikely to be chosen as play partners and have difficulty getting others to play. These sanctions might influence an individual's reproductive fitness. If a dog doesn't want to play, then she shouldn't bow.

PROMOTING EGALITARIANISM AND REDUCING INEQUITIES

Dogs, wolves, coyotes, and other animals engage in role reversing and self-handicapping to maintain social play. Each of these strategies helps to reduce inequalities in size and dominance rank between players and to promote the reciprocity and cooperation that's needed for play to occur. Given that play has to be cooperative and carefully negotiated, any action that can reduce inequities and foster symmetry would be well used during social play so that the interaction isn't terminated.

Self-handicapping (or "play inhibition") happens when an individual performs a behavior pattern that might compromise her outside of play.

For example, a coyote might decide not to bite her play partner as hard as she can, or she might not play as vigorously as she can. Inhibiting the intensity of a bite during play helps to maintain the play mood. The fur of young coyotes is very thin and an intense bite results in high-pitched squeals and much pain to the recipient. An intense bite is a play-stopper. In adult wolves, a bite can generate as much as 1,500 pounds of pressure per square inch, so there's a good reason to inhibit its force.

Role reversing happens when a dominant animal performs an action during play that wouldn't normally occur during real aggression. For example, a dominant wolf wouldn't rollover on his back during fighting, but would do so while playing, making himself more vulnerable to attack. In some instances, role reversing and self-handicapping might occur together. A dominant wolf might roll over while playing with a subordinate dog and at the same time inhibit the intensity of a bite. Self-handicapping and role reversing, similar to using specific play-invitation signals, might indicate an individual's intention to continue to play and seem to be important in maintaining fair play.

Although we've focused on dogs and their wild relatives, other animals also work hard to negotiate fair play. For example, Australian biologists Duncan Watson and David Croft observed red-necked wallabies engaging in self-handicapping. These playful creatures adjust their play to the age of their partner. When a partner is younger, the older animal adopts a defensive, flat-footed posture, and pawing rather than sparring occurs. The older player is also more tolerant of its partner's tactics and takes the initiative in prolonging interactions.

Sergio Pellis discovered that sequences of rat play consist of individuals assessing and monitoring one another and then fine tuning and changing their own behavior to maintain the play mood. When the rules of play are violated, when fairness breaks down, so does play. Even in rats, fairness and trust are important in the dynamics of playful interactions. Pellis observed that when adult rats play, subordinate individuals direct more playful contacts (touching or nearly touching a second rat's nape with snout) at the dominant rat and they try to retain a symmetrical play relationship so that they aren't injured and the dominant rat knows that they're playing and not fighting. Dominant rats tend to evade these encounters with adult defense tactics, while subordinate rats, when playfully attacked, roll over into the juvenile defense position. The initiation of such playful attacks by the subordinate rat may lead the dominant rat to tolerate the subordinate's presence.

So, why do animals carefully use play signals to tell others that they truly want to play and not beat them up, why do they engage in self-handicapping and role reversing, why do they fine tune play to keep play going while having fun on the run? Well, it's plausible that during social play, while individuals are having fun in a relatively safe environment, they learn ground rules about what behavior patterns are acceptable to others, for example how hard they can bite, how roughly they can interact, and how to resolve conflicts without having to stop the playful encounter.

There's a premium on playing fairly and trusting others to do so as well. It's also possible that individuals might generalize codes of conduct learned in play with specific individuals to other group members and to other situations where justice might come into play, such as reciprocity in grooming, sharing food, negotiating social status, and defending resources. There are codes of social conduct that regulate actions that are and aren't permissible, and the existence of these codes has much to say about the evolution of social morality. What could be a better atmosphere in which to learn about the social skills underlying fairness and cooperation than during social play, where there are few penalties for transgressions?

THE PLEASURE OF PLAY

In his book *The Descent of Man, and Selection in Relation to Sex* Charles Darwin wrote, "Happiness is never better exhibited than by young animals, such as puppies, kittens, lambs, &c., when playing together, like our own children." Animals typically only play when they're relaxed, unstressed, and healthy, so the inherent joy and serenity in play often spreads to anyone who is watching.

Ethologist Jonathan Balcombe says that pleasure is "one of the blessings of evolution." It's one of the ways in which nature rewards adaptive behavior. Humans (especially the Puritans among us) may think that morality and pleasure are opposing forces; anything fun is undoubtedly also naughty. Yet nature knows better. Balcombe notes, "sensory plea-sure induces behaviors that improve homeostasis," presumably by helping to maintain and by rewarding behaviors that improve social homeostasis. Joy (or, in stuffy scientific terms, "positive affect") and pleasure play a key role in morality.

What we can see with our eyes is also being borne out by scientific research. Studies of brain chemistry in rats support the idea that play is pleasurable and fun. Renowned neurobiologist Jaak Panksepp discov-ered in rats that an increase in opioid activity may enhance the pleasures and rewards associated with playing. If this is true in rats, and we already know it's true in humans, then there's little reason to believe that the neurochemical basis of play-inspired joy in dogs, cats, horses, and bears would differ substantially.

APOLOGIZING AND FORGIVING: HOLDING GRUDGES IS A WASTE OF TIME

What about forgiveness? This is another moral sense that is often attrib-uted solely to humans, but the renowned evolutionary biologist David Sloan Wilson argues that forgiveness is a complex biological adaptation. In his book *Darwin's Cathedral: Evolution, Religion, and the Nature of Society*, Wilson writes, "Forgiveness has a biological foundation that extends throughout the animal kingdom." And further, "Forgiveness has many faces—*and needs to*—in order to function adaptively in so many different contexts." While Wilson concentrates mainly on human societies, his views can be easily and legitimately extended to nonhuman animals. Indeed, Wilson points out that adaptive traits such as forgiveness might not require as much brain power as once thought. This isn't to say that animals aren't smart, but rather that forgiveness might be a trait that is basic to many animals even if they don't have especially big and active brains.

Play sequences often involve acts of forgiveness and apology. For example, if Jethro bit Zeke too hard, and play stopped for a moment, Jethro would then bow and tell Zeke by bowing that he didn't mean to bite Zeke as vigorously as he did. Jethro is asking for forgiveness by apologizing. In order for play to ensue, Zeke has to trust that Jethro meant what he said when he bowed, that Jethro was being honest. While this may seem farfetched to some readers, the facts show that play bows are used strategically to maintain the play mood when it might otherwise end.

So, all in all, social play is a perfect activity in which to look for moral behavior in animals (and in humans). The basic rules of the game are: ask first, be honest and follow the rules, and admit when you're wrong.

INEQUITY AVERSION: I'LL HAVE WHAT SHE'S HAVING

An additional area of research sheds light on animals' sense of fairness and equity. Several primate studies have focused attention on "inequity aversion," a negative reaction arising when expectations about the fair distribution of rewards have been violated. There are thought to be two basic forms of inequity aversion: the first is an aversion to seeing another individual receive more than you do; the second is an aversion to receiving more yourself than another individual receives. Only the first type of inequity aversion—the "That's not fair, she got more than I did" variety—has been explored in animals.

Sarah Brosnan and Frans de Waal tested five female captive capuchin monkeys for inequity aversion. Capuchin monkeys are a highly social and cooperative species in which food sharing is common; the monkeys carefully monitor equity and fair treatment among peers. Social monitoring for equity is especially evident among females. Brosnan and de Waal note, "Females pay closer attention than males to the value of exchanged goods and services."

Brosnan first trained a group of capuchins to use small pieces of rock as tokens of exchange for food. Pairs of females were then asked to barter for treats. One monkey was asked to swap a piece of granite for a grape. A second monkey, who had just witnessed the rock-for-grape trade, was asked to swap a rock for a piece of cucumber, a much less desirable treat. The short-changed monkey would refuse to cooperate with the researchers and wouldn't eat the cucumber and often threw it back at the human. In a nutshell, the capuchins expected to be treated fairly. They seemed to measure and compare rewards in relation to those around them. A single monkey who traded a rock for a cucumber would be delighted with the outcome. It was only when others seemed to get something better that the cucumber suddenly became undesirable.

Skeptics have argued that these monkeys are not exhibiting a sense of equity, but rather a sense of greed and envy. And indeed they are. But greed and envy exist as counterparts to justice; unless you feel short-changed, why would you feel envious? And why would you feel short-changed unless you thought you deserved more?

Brosnan and de Waal speculate that monkeys, like humans, are guided by social emotions or "passions" that modulate an individual's response to "the efforts, gains, losses and attitudes of others." Passions such as gratitude and indignation have evolved to nurture long-term coopera-tion, and seem to exist in monkeys as well as in humans, and they may exist in other species.

Of these passions, the one that jumps out at anyone reading Brosnan and de Waal's study is indignation, because it smacks of strong anthro-pomorphism. Indignation is the emotion aroused by a perceived sense of injustice. As de Waal notes in *Good Natured*, "the outraged reaction that [injustice] may trigger serves to clarify that altruism is not unlim-ited, it is bound by the rules of mutual obligation" (i.e., fairness). De Waal also considers gratitude. In a 2005 *Scientific American* essay about reciprocity in monkeys, he writes, "This reciprocity mechanism requires memory of previous events as well as the coloring of memory such that it induces friendly behavior. In our own species, this coloring process is known as 'gratitude,' and there is no reason to call it something else in chimpanzees." De Waal clearly recognizes the implications of these observations of monkeys when he claims, "Thus, reading *A Theory of Justice*, an influential book by the contemporary philosopher John Rawls, I cannot escape the feeling that rather than describing a human innova-tion, it elaborates on ancient themes, many of which are recognizable in our nearest relatives."

Another study by Brosnan, de Waal, and Hillary Schiff suggests that chimpanzees also display a sense of inequity aversion. As with the ca-puchins, chimpanzees in a similar experimental setup showed negative reactions to inequity in reward. This study went a bit further than the capuchin study, and made initial forays into some fascinating nuances of fairness behavior. Although chimpanzees responded to discrepancies in level of reward, they seemed indifferent to discrepancies in level of effort. Like the capuchins, the chimpanzees did not seem bothered when they received a superior reward (they didn't show the second form of inequity aversion). Also, the strength of chimps' reactions to inequity varied according to social context, including group size and relatedness. In long-term and tightly knit social groups, the chimpanzees showed more tolerance for inequity. Perhaps this is because individuals keep track of who does what to whom and, as predicted by renowned evolutionary biologist Robert Trivers in his theory of reciprocal altruism, we would expect such patterns of social behavior to arise in long-lived groups in which individuals recognize one another over time. It's important that individuals remember who did what to whom and who should preferentially be repaid in the future.

These studies suggest that justice is situational. What's acceptable in one social context might be unacceptable in another. So, in order to learn more about justice in animals, we need to take into account the specific context in which behaviors are expressed, for example, the size of the group, the longevity of social relationships, and the stability of group membership, which is related to nonsocial environmental conditions. One shoe doesn't fit all.

FAIRNESS AND FITNESS: THE PENALTIES OF BREACHING TRUST

One big question of interest to biologists is how differences in the performance of a given behavior influence an individual's reproductive success. Ethologist Niko Tinbergen among others noted that making this connection should be one of the goals of behavioral research. So, might differences in play and variations in fair play affect an individual's reproductive fitness? It's almost impossible to directly link fair play with an individual's reproductive success or fitness. But some intriguing data from coyotes speak to the relationship between play and fitness.

Coyotes are fast learners when it comes to fair play, as they should be, for there are serious penalties when they breach the trust of their friends. Biologists call these penalties "costs," which means that an individual might suffer some decline in his or her reproductive fitness if he or she doesn't play by the expected rules of the game. Fieldwork on coyotes has revealed one direct and immediate cost paid by individuals who fail to engage in fair play or who don't play much at all: youngsters who don't play as much as others, either because they are avoided by others or because they themselves avoid others, are less tightly bonded to members of their group. These individuals are more likely to leave their group and try to make it on their own. But life outside the group is much riskier than within it. In their seven-year study of coyotes living in the Grand Teton National Park outside Moose, Wyoming, Marc and his colleagues found that about 60 percent of yearlings who drifted away from their social group died, whereas fewer than 20 percent of their stay-at-home peers did. Was it because of play? We're not sure. The detailed information that's needed to know for certain is impossible to collect in the field. However, data collected on captive coyotes show that individuals who don't play fairly play less frequently than fair players, and the lack of play is a major factor in individuals spending more time alone, away from their littermates and other group members.

What about humans? All of these tantalizing threads mirror what we know of human responses to inequity. For example, we know that people who feel they're being treated unfairly have a higher risk of developing heart disease. Researchers have speculated that feeling slighted might prompt biochemical changes in the body because of the negative emotions associated with being treated unfairly. Thus, the positive emotions associated with a feeling of being treated fairly likely have deep-seated evolutionary roots. Along these lines, medical epidemiologist Richard Wilkinson notes in his book *Unhealthy Societies: The Afflictions of Inequality* that the most egalitarian countries, such as Norway, tend to have healthier populations than countries in which there are large disparities between rich and poor, such as the United States. He speculates that inequality leads to ill health because of the physiological consequences of social stress.

FAIRNESS, TRUST, AND SELF-INTEREST

Primatologist Robert Sussman and ethicist Audrey Chapman note that group living in animals involves compromising individual freedoms, and that these compromises can go against self-interest. Moving beyond self-interest in turn seems to involve trusting others within one's social network. Corporate lawyer Lawrence Mitchell, writing in *Stacked Deck* about selfishness and trust in America, says something remarkably similar, and raises some points that are worth considering in our discussion of justice in animal societies. Our comments on Mitchell's ideas are necessarily speculative because there are extremely few data that bear on the question of justice in animals. However, we hope this discussion will stimulate further research.

To quote Mitchell, a "society of self-interest makes trust difficult if not impossible . . . [i]t is an ethic that cannot sustain trust. Because it cannot sustain trust, it creates relationships of mutual suspicion and self-protection. It makes it far more difficult to have meaningful and rich interactions with people, at least those outside our immediate families and close circles of friends (and we may be forgiven for being wary even in these relationships)." Mitchell argues that in human societies unfairness breeds mistrust, and mistrust creates social instability. Is it farfetched to wonder whether the integrity and efficiency of a pack of wolves, a pride of lions, a herd of elephants, or a troop of chimpanzees rests on individuals trusting the intentions of other group members? No. Trust is essential for maintaining group cohesion. It is important in social play and in reciprocity, both of which foster group living.

Mitchell also argues that fairness is deeply rooted in vulnerability; vulnerability is a normal human condition; we are all vulnerable. "We can start by changing our minds—by changing the ways we think about these issues. We can start by understanding that fairness is all about vulnerability. If we do, we will breed trust. We will breed social cohesion. We will build community." Are social animals vulnerable in similar ways? We think they are, and that understanding the vulnerability of social animals will help us understand more about wild justice.

PHILOSOPHIZING JUSTICE: JUSTICE ISN'T SIMPLY AN ABSTRACT PRINCIPLE

Of the three clusters, justice is the one most likely to raise eyebrows. It sounds funny to say that animals can behave justly. This is primarily a reaction to the way justice has been framed in our cultural discussion. It is generally spoken of as a set of abstract principles about who deserves what. And animals, as far as we know, don't think in abstractions.

But as we suggested in the first chapter, morality—including justice—is really not about abstractions, at least not primarily. Robert Solomon writes, in A Passion for Justice, "Justice presumes a personal concern for others. It is first of all a sense, not a rational or social construction, and I want to argue that this sense is, in an important sense, natural." Solomon's point is reflected in our everyday use of language: we often use the phrase "sense of justice." This suggests that justice, like empathy, is a sentiment or a feeling, and not only or even primarily an abstract set of principles.

Paul Shapiro makes a similar point in his essay "Moral Agency in Other Animals." He writes, "Being able to care about the interests of others is central to what matters in morality, and arguably more important than abstract principles regarding proper conduct." Caring about the interests of others, and comparing these interests to your own, is the essence of justice.

Frans de Waal, who is typically quite generous in ascribing moral behaviors to animals, is circumspect about justice. When asked in an interview in Believer magazine whether animals have a sense of fairness, he equivocates. He admits that animals have moral emotions, including empathy. But, he says, "to get to morality you need more than just the emotions . . . You need to be able to look at a situation, and make a judgment about that situation even though it doesn't affect you yourself." You need a kind of distance. You need to be able to play the role of what philosophers call the "impartial spectator" and make moral judgments about situations that don't directly affect you. In chimpanzees, de Waal says, you won't find a concept of fairness about interactions among others.

De Waal's comments remind us of an important truth: human morality is unique. In human societies, the capacity to think abstractly about who deserves what and why is vitally important. We might view this as a human innovation—a specialization or refinement of the capacity for justice. Justice as expressed in human societies is arguably more complex and more nuanced than in other animal societies. But this in no way suggests that animals cannot and do not also have a sense of fairness.

Skeptics, particularly after reading de Waal's comments, might object that animals can't have a sense of justice because they can't be impartial. Impartiality is a principle of justice holding that decisions about who gets what are made without bias, without prejudice against race or sexual orientation, and without nepotism or other inappropriate preferences. Justice, the saying goes, must be blind. Although impartiality functions as an important principle in certain contexts in which justice is in play, these contexts are limited in number and scope and encompass only a small corner of fairness and justice in human social encounters. So, whether or not animals can be impartial (which, incidentally, has never been studied) is really irrelevant to whether they have a sense of justice and fairness.

CONNECTING THE CLUSTERS

As we draw our discussion of the three clusters to a close, it is worth giving some thought to how the various threads of moral behavior in animals connect and overlap. These are just tentative observations, based on the limited data available at this point in time. As research delves more deeply into these three behavioral clusters, the interconnections will certainly become more distinct and more robust.

Our informed guess would be that of the three clusters, justice represents the most highly developed and evolved set of behaviors, requiring the most neural complexity and nuanced emotional sensitivity. Justice probably rests on the foundation of both empathy and cooperation, and will be more narrowly distributed than the other clusters of behaviors.

Fairness is closely tied to cooperation, particularly more complex forms of cooperation such as reciprocity. Some of the basic behavioral elements of cooperation are necessary for justice. For example, in cooperative relationships, it is important to be able to compare your own effort or contribution with that of others, and there needs to be parity in contribution (parity in both cost and benefit). This capacity to compare, which is cognitively complex and requires memory of past encounters, expectations about the future, and a nuanced assessment of another animal's character, is also at the heart of justice.

Trust, which is essential to cooperative and reciprocal exchanges, is also a basic element of fairness, particularly in the context of social play. The justice and cooperation clusters both include behaviors related to punishment of cheaters, free riders, and liars, including negative emotions that arise when expectations are not met. Our informed guess would be that justice and a sense of fairness have evolved out of the more basic repertoire of cooperative and altruistic behavior. As the renowned neuroscientist Antonio Damasio has argued, "It is not difficult to imagine the emergence of justice and honor out of the practices of cooperation."

We believe that justice is also rooted in empathy. A sense of fairness clearly requires the capacity to read the intentions and emotional states of others, as do complex forms of cooperation. Recall the discussion of play behavior as a constant stream of subtle communications about intentions, beliefs, and desires.

It is possible that research in neuroscience will help elucidate the connections between justice and empathy. Neuroscientists have begun investigating the neural foundations of both empathy and fairness, and some intriguing connections seem to be emerging. A study published in Nature by neuroscientist Tania Singer and her colleagues showed that people feel empathy toward those who have treated them fairly in social interactions. But this empathic response is not activated or is activated much less strongly toward people who have been unfair. This suggests a close neurological tie between empathy and justice, almost certainly in humans but perhaps also in other species. Justice might also be mediated by mirror neurons. We've noted that mirror neurons may be involved in the sharing of play intentions; also that play is contagious. These intriguing connections call out for further study.

It's likely that altruism and empathy are also intimately linked, both in terms of their evolutionary history and their proximate mechanisms. Social psychologist Daniel Batson has proposed that the empathic response is one of the central mechanisms underlying altruistic behavior. There is considerable support among psychologists for what Batson termed his "empathy-altruism hypothesis." Whether or not empathy and altruism are similarly linked in animals remains an open question, but parsimony would suggest an affirmative answer. And when we think back to the research on empathy in animals, we can see that in many instances, the behaviors we observe belong also within the cooperation cluster. Recall

Iain Douglas-Hamilton's story of Grace; the elephants in her herd not only seemed to empathize with her suffering, but they tried to help. The story of mice in the sink, too, was not just about recognizing the plight of another animal, but finding a way to relieve its suffering.

The three clusters of behavior seem clearly to weave together into an integrated whole, like different colors and textures of thread in a magnificent tapestry. New research will continue to fill in detail and add depth and nuance to the picture.

WHERE TO NOW?

Throughout these pages, you've probably found yourself pondering questions that aren't really scientific, but rather more philosophical in nature. If animals really do have morality, how would this change our understanding of ethics within our own species? If morality comes "from nature," as it were, does this make morality somehow less real or less binding? What about those who argue that morality is grounded in religious belief? Do animals have religion, too? And aren't there important differences between our own systems of morality and those found in animal societies?

In the first five chapters, we've tried to keep our attention focused primarily on the scientific data supporting our hypothesis that animals have morality. But different kinds of questions—philosophical ones—have been looming in the background, rather like an elephant in the room that we have thus far refused to acknowledge. We have kept the elephant obscure on purpose, so that we could really focus on what the available data suggest about moral behavior in animals. But these other questions, these philosophical concerns, are vitally important, too. And we turn our attention now to addressing some of the philosophical implications of Wild Justice.

6 ANIMAL MORALITY AND ITS DISCONTENTS

A NEW SYNTHESIS

As we promised in our preface, the proper treatment of wild justice would take us on a journey over hills, into valleys, and around turns. And it did, taking even more turns then we anticipated. So how can we tie it altogether?

THE CASE OF THE MIDWIFE BAT

Renowned bat biologist Thomas Kunz of Boston University and his research team made a startling discovery that was important enough to be published in the prestigious *Journal of Zoology*. While they were observing a captive colony of Rodrigues fruit-eating bats in Gainesville, Florida, they saw a female help another female to give birth. The pregnant female was hanging in the typical perching position, head down, feet up. But when bats give birth they switch to a head up, feet down position, so the midwife perched in front of the pregnant female in the correct birthing position as if to show her how to do it, to tutor her so she could give birth. The pregnant female copied her. The helping female then licked the mother's anogenital region, and when the baby emerged she groomed it and helped it crawl towards the mother's nipples so it could feed. Kunz concluded that "such cooperative behaviour is probably common in bats that roost in colonies" but very few people have actually seen secretive bats give birth. Furthermore, the midwife bat wasn't related to the new mother. So why did she help her, what did she get out of it, and did she understand that the mother needed help? Was she simply doing what's right, the moral thing to do? Was she a chiropteran philosopher?

Wild justice raises a number of philosophical questions about how we should understand morality and how we should understand animals. By *philosophical*, we mean simply the quest for a deeper understanding of "big" questions about the nature of reality and the proper conduct of life. We are particularly interested in *moral* philosophy, which refers to a philosophical investigation of the moral domain. Science is only part of the picture, particularly when it comes to questions about right and wrong, good and evil, and the meaning of life in general.

We don't intend this chapter to be a thorough discussion of the philosophical implications of animal morality, for this isn't our agenda. We simply want to outline what we see as the most pressing and interesting questions, and where further conversation in ivory towers, at professional conferences, in coffee shops, and at dog parks will be most relevant.

One unifying thread is our continued appeals to the importance of evolutionary continuity between humans and other animals leading to the conclusion that we're not the sole occupants of the moral arena. In suggesting that there's continuity even in moral behavior, wild justice seems to jeopardize the special status of humans as separate from and above the rest of nature. This, in turn, seems to threaten an ideal of human dignity and right. Wild justice also raises questions about the mixing of biology and ethics, the mixing of facts and values, as scholars put it, and what this means about how and even if evolutionary theory is applied to patterns of human social behavior. Should morality be wrested from the hands of the humanities and reside solely in the province of biology? In drawing a picture of animals as beings with rich cognitive, emotional, and social lives, wild justice invites a serious reconsideration of the uses to which we put animals in research, education, and for clothes and food, among other things.

Some of the most important philosophical questions arise in relation to the definition of morality and the extension of this term to include nonhuman animals. In including some animals within the sphere of morality, we force a reconsideration of what have been assumed to be the essential ingredients of morality, elements such as reflective judgment, agency, and conscience. *Wild Justice* suggests a picture of morality in which agency and conscience are only pieces of a much bigger and more interesting picture. The meaning of morality needs to evolve in light of a broad and interdisciplinary perspective and newly emerging research. In addition to deciding which capacities are necessary ingredients of morality, we also need to know whether and in what sense animals can be said to possess these capacities.

ON THE MEANING OF MORALITY: FINESSING THE BEAST

We've defined morality in broad terms as a suite of other-regarding behaviors falling into the three rough clusters of cooperation, empathy, and justice. The definition is broad enough that the behavior of many social animals, not only *Homo sapiens*, falls within its boundaries. Have we watered down its meaning by defining the word so expansively? Think back to the empathy chapter, and to the way in which Stephanie Preston and Frans de Waal define empathy as a whole spectrum of behavior patterns that share the common feature of emotional linkage. Rather than diluting the concept, their refining of empathy makes it more specific, more detailed, and more meaningful. The same thing happens with morality; it's not a unitary phenomenon, so giving a broad description that encompasses its diversity and range is going to give it more meaning, not less. Morality is a spectrum of behaviors that share the common feature of concern about the welfare of others. We've put down a trawling net, and we've caught all sorts of things. And this is really what morality is like: you can't just put down a small net and expect to catch your prey. Morality is not like a little minnow but rather a whole teeming sea of creatures.

People who object to our definition of morality are likely reacting not to the broadness of our definition, per se, but to the philosophical implications of defining morality so that it includes nonhuman animals. The idea that morality is not unique to humans will be, for many people, not only surprising but also suspicious, because it seems to challenge many assumptions about what makes humans special.

MARKING DIFFERENCES IN KIND
AND DIFFERENCES IN DEGREE

We've emphasized evolutionary continuity throughout the book and have argued that humans share with other social mammals the same basic suite of moral behaviors, namely fairness, cooperation, and empathy. We've also suggested that morality may exist along a continuum, from simpler to more complex patterns of behavior. Morality can be thought of as nested levels of increasing complexity and specificity much like Preston and de Waal's Russian-nested-doll model of empathy. It is likely that humans share with other social mammals some of the inner layers of moral behavior. But humans also appear to have evolved an unusually high level of moral complexity.

Yet just how different are humans and animals? This question is often answered with language from Darwin; animals and humans are either different in kind or they are different in degree. The theory of evolution would seem to answer the question with "by degree." All mammals, for example, share a common origin and have gradually differentiated in response to environmental pressures. Yet even people who generally espouse an evolutionary view have tended to exclude animals from the moral domain. When it comes to morality, humans have long been considered different in kind and not merely degree. This has been cashed out as "we" have it and "they" don't. This narrow-minded assumption obviously needs to be reconsidered, and we have argued throughout *Wild Justice* that animal morality is different in degree, not in kind, from human morality. While animals can surely be booted out of the moral arena by narrowly defining morality, a species-relative approach such as the one we take begins with a broad and inclusive definition and works from there to elucidate patterns of moral behavior unique to each species.

So what patterns are unique to humans? Harvard University philosopher Christine Korsgaard argues that the capacity to assess and adopt intentions and to make judgments about whether a particular course of action is morally justified are unique to humans and represent a break from our animal past. The human prefrontal cortex, the area of the brain responsible for judgment and rational thought, is more highly developed in humans than in other animals. With judgment and rational thought (what is often called *reason*), we gain self-consciousness about the grounds of our actions and gain a corresponding capacity for self-governance and conscious control. We are conscious of the *grounds* of our beliefs and actions, and this consciousness is the source of reason, a capacity distinct from intelligence. "[T]he capacity for normative self-government and the deeper level of intentional control that goes with it is probably unique to human beings. And it is in the proper use of this capacity—the ability to form and act on judgments of what we ought to do—that the essence of morality lies, not in altruism or the pursuit of the greater good." It is because animals lack this capacity for reflective self-control that we don't hold them responsible. We don't hold them morally culpable for following their strongest impulses.

Humans also use language to articulate and enforce moral norms, another potential difference in kind. As Robin Dunbar's work on gossip and reputation suggests, language and morality are intimately tied. Dunbar, who works at the Institute of Cognitive and Evolutionary Anthropology at the University of Oxford in the UK, argues that language has been evaluative from its origins; it has been used to communicate socially important information about each other, such as who is trustworthy and who will reciprocate. Our words express anger, contempt, and approval in our public utterances. But does language separate humans from other animals? Anthropologist Terrence Deacon thinks it does. In his book *The Symbolic Species*, Deacon argues that although there is undoubtedly an unbroken continuity between human and nonhuman minds, there is also a singular discontinuity: humans use language to communicate. The use of words has changed our brain over time. Deacon notes, "[T]he first use of symbolic reference by some distant ancestors changed how natural selection processes have affected hominid brain evolution ever since." If our brains are significantly different, and morality is essentially a product of the brain, then wouldn't we possibly be unique in this respect? Animals communicate about morality, but not with language. This would be an important subject for comparative work.

Even if there are bona fide differences in kind, this does not mean that many aspects of morality aren't also shared, or that there aren't significant areas of continuity and overlap. We view each of these possibly unique capacities (language, judgment) as outer layers of the Russian doll, relatively late evolutionary additions to the suite of moral behaviors. And although each of these capacities may make human morality unique, they are all grounded in a much deeper, broader, and evolutionarily more ancient layer of moral behaviors that we share with other animals.

UNIQUENESS

Many people are uncomfortable with the idea of ascribing morality to animals because it seems to threaten the uniqueness of humans. This concern may take any number of forms. For example, many Christian theologians see a sharp line between humans (who are created in the image of God) and the rest of creation (who are not), and it is a matter of theological importance to respect this doctrinal distinction. Many philosophers believe that human uniqueness provides the fundamental grounding of human dignity, and thus serves to protect classes of humans (such as fetuses and the severely mentally handicapped) who might otherwise be treated as less than human (in other words, like animals). Some might also consider it important to maintain a sense of human uniqueness, or animal difference, because it serves as ethical justification for the use of animals in scientific research.

We offer two brief comments to put these concerns to rest. First, the idea that ascribing morality to animals could lead to loss of respect for vulnerable or "marginal" people is just bizarre—it follows no clear logical path. The presence of morality in animals does not threaten human uniqueness, nor does it threaten vulnerable populations. In fact, comparative work of the sort we suggest serves to clarify and illuminate our uniqueness. Although there is evolutionary continuity, there is also difference. And we can be very clear about what this difference is. And we can celebrate this difference, feel proud of it, and use it to ground a principle of human dignity.

Second, human uniqueness, such as it is, cannot logically serve as ethical justification for the instrumental use of animals. There is, again, no clear logical path in this direction. Of course, human uniqueness has been used in this way. But this doesn't make it logical. Human dignity does not carry as its correlate the indignity of animals.

Uniqueness is something to be celebrated, and can be a tool for gaining a deeper understanding of and empathy toward others. Each species has its own unique features that make it beautifully adapted and fascinating to study. And just as each individual human is unique, with unique physical features, personality, and life experiences, so is each individual animal. As any dog or cat owner knows, there is a great deal of individuality from one animal to the next. There is significant behavioral and dispositional variation among individuals within a species—what we might call personality. All brown-eared bats may look alike to us, but to them, each individual is unique. We need to keep this individual variation in mind during ethological research on animals and when thinking about animal welfare—about what, for example, animals need to be happy and healthy.

DO ANIMALS HAVE WHAT IT TAKES TO BE MORAL BEINGS?

There are a number of possible objections to wild justice that center on the issue of whether animals have one or another particular skill or capacity that is considered a necessary ingredient of morality. Here are a few examples:

- Animals aren't smart enough to have morality
- Animals don't have moral emotions, and thus lack morality
- Animals can't empathize, and thus can't be moral
- Animals aren't rational, and thus lack morality
- Animals lack reflective judgment, and thus lack morality
- Animals aren't moral agents, and thus lack morality
- Animals lack conscience, and thus lack morality

We've already discussed and jettisoned the first four objections. Animals clearly have the cognitive and emotional capacities for moral behavior and display empathy and rational thought. We've also discussed the fifth objection, reflective judgment, and argued that while this may be a bona fide difference between animals and humans, reflective judgment isn't a precondition for moral behavior.

Let's now consider the final two objections, agency and conscience. At first glance these may seem like scientific objections, and, in fact, in the philosophical literature they're often framed as such. In each case, our attention is drawn to a statement about what animals are like. They lack agency; they have no sense of conscience. "Yes," you might think, "that is certainly true." But notice that each of these objections also contains an implicit assertion about what morality itself is like, that agency and conscience are essential components of morality.

Claims about what morality entails also need some attention because some of them (or perhaps all of them) may turn out to be wrong. Questions about the defining characteristics of morality are both scientific and deeply philosophical and spiritual. Although the prominent biologist E. O. Wilson famously argued that "scientists and humanists should consider together the possibility that the time has come for ethics to be removed temporarily from the hands of the philosophers and biologicized," we, a scientist and a philosopher, argue that morality is not and should not be the peculiar province of biology.

The concepts of agency and conscience have no straightforward scientific definition. These are essentially philosophical concepts and, as such, their meaning is open to debate and disagreement. Their application to animals, in particular, is open to debate; animals cannot be categorically excluded. Because there is controversy, we might usefully rephrase the naïve statements with questions: to what degree does conscience play a role in moral behavior, what kinds of agency do animals seem to exhibit, and are these relevant to moral behavior? Now we have some interesting and important questions that can guide future research into and discussion of animal and human morality.

MORAL AGENCY: ARE ANIMALS RESPONSIBLE FOR WHAT THEY DO?

Whenever either of us speaks about morality in animals, one of the first questions raised by our audience is whether animals have agency. Agency is a philosophical concept that means the capacity to act freely or, in philosophical parlance, to act autonomously. A person is thought to be a moral agent when she or he freely chooses to act one way rather than another in response to a moral dilemma. By claiming that animals have morality, many people assume that we're also claiming that animals are moral agents.

Animals, at least according to most Western philosophical accounts, cannot be moral agents because they are guided solely by instinct. However, it's not true that animals act solely on instinct, so we need some concept of moral agency in animals. Nor is it true that humans act "autonomously" as this has been generally understood, so ideas of human agency also need to evolve in light of new research in neuroscience and cognitive psychology.

Because agency has been a garden-variety justification for excluding animals from moral consideration, we need to approach the concept with caution. Agency, in general, needs to be rethought. Even those who accept that animals display some moral behaviors may have trouble going a step further and believing that animals can be called moral agents. We suggest that this next step is neither very large nor particularly troubling. In philosophical lingo, a moral agent is one who freely chooses to act in one way rather than another and is held responsible for his or her actions. *Moral agent* is typically contrasted with *moral patient*, and this agent/patient dyad is used to draw a distinction between those who can make moral choices and those who cannot, as a way of ascribing responsibility for actions or omissions. Animals, human infants, and humans with severe cognitive impairments, for example, have typically been categorized as patients, individuals who are incapable of being responsible for making moral choices. This dichotomy between agent and patient, though perhaps useful in limited contexts, can also be misleading.

To claim that animals have moral agency is not, of course, to argue for sameness. Paul Shapiro correctly notes, "It would be naïve to assert that other animals are moral agents *in the same sense* in which most adult humans are." Moral agency is species-specific and context-specific. Furthermore, animals are moral agents *within the limited context of their own communities.* They have the capacity to shape their behavioral responses to each other based on an emotionally and cognitively rich interpretation of a particular social interaction. Wolf morality reflects a code of conduct that guides the behavior of wolves within a given community of wolves.

Wolves are agents only within this context. The predatory behavior of a wolf toward an elk is *amoral*—it is not subject to condemnation or accolades.

Animals actively make choices in their social encounters, including whether or not to help others. Stanley Wechkin's monkeys, Russell Church's rats, and Christine Drea's hyenas all made a choice to pull or not to pull a rope, to push or not to push a lever, to help or not to help. So too did Tom Kunz's midwife bat make a choice to help a mother in distress. Where there's flexibility and plasticity in behavior, there's choice, there's agency. This is the very reason we do not include insects among our moral animals, because as far as we know their behavioral patterns are rigid; they don't seem to "choose" in the same sense that the social mammals choose. And this is why we set threshold requirements for our moral animals: flexibility, plasticity, emotional complexity, and a particular set of cognitive skills.

Even behavior that is conditioned or instinctual can count as moral. Indeed, research suggests that a good deal of human morality is conditioned and instinctive. It wouldn't make sense to say that humans are only moral agents in those rare circumstances in which they act on a moral abstraction. And we should remember that parents and teachers go to great lengths to *condition* children to behave in morally appropriate ways.

Although we are willing to call animals moral agents, we believe that the language of agent and patient is likely to promote philosophical confusion and should ultimately be avoided.

DARWIN'S DOGS: CONSCIENCE AS A MORAL COMPASS

Charles Darwin proposed that any animal whatever, endowed with "well-marked social instincts," could develop of sense of conscience. In *The Descent of Man*, Darwin wrote, "Besides love and sympathy, animals exhibit other qualities connected with the social instincts, which in us would be called moral; and I agree with Agassiz that dogs possess something very much like a conscience." Darwin believed that animals possess the "power of self-command" in that they are capable choosing one course of action over another. He also pointed out that on occasion there would be an internal struggle over competing impulses. Darwin described conscience as an "inward monitor" that tells an animal that it would be better to follow one impulse rather than another. Dogs will, for example, refrain from stealing food off the counter, even when their master is not present. "The one course ought to have been followed, and the other ought not; the one would have been right and the other wrong."

Impulse control is certainly an important component of moral behavior, and moral psychologists such as Lawrence Kohlberg have long argued that the development of impulse control in young children is important in the development of mature morality. It is also clear that the animals in our moral taxonomy are capable of impulse control. Yet it isn't clear that impulse control and conscience are equivalent, or that impulse control alone is sufficient for mature moral behavior. Nor it is clear whether animals other than humans have conscience.

Writing about humans, the distinguished ethologist Robert Hinde argues in his book *Why Good Is Good* that having a "good conscience" implies the maintenance of congruency between one's action and what he calls the "self-system," the internalized moral norms of a given society. "Moral judgments," he says, "depend on comparisons between values incorporated in the self-system and observed or intended action." At least some species of animals (our moral animals) have "conscience" in the sense that moral rules are internalized, and some self-monitoring of behavior occurs.

On the other hand, conscience may be something more particular than either impulse control or the internalization of a set of norms about good and bad. Anthropologist Christopher Boehm's work on social sanctioning and the origins of conscience in humans suggests a somewhat different answer to the question of conscience in animals. According to Boehm moral conscience is a uniquely human capacity and conscience is an essential component of morality. His hypothesis is that conscience evolved in *Homo sapiens* in response to the shift during the Middle to Late Pleistocene from subsistence to hunting of large game. Boehm also claims that hunting large game required intense cooperation and that fair distribution of meat among group members was enforced though systems of social sanctioning. As Boehm says, "bands had to gang up physically against their alphas to ensure efficient meat distribution—thereby inventing a systematic and decisive type of group social control." This, he continues, "set the stage for morality to develop, as a new, more socially-sensitive type of personal self-control became adaptive for individuals living in these punitive groups." So conscience first, then morality.

Boehm goes on to argue, "It is having a self-judgmental conscience that makes us distinctively 'moral,' and this brings us to shame as a specialized manifestation of conscience. Chimpanzees and bonobos experience neither socially-induced facial flushing nor any other overt evidence of feeling shame, so there is no obvious preadaptation for specifically moral emotions." Neuroscientist Antonio Damasio's work confirms that conscience is particularly well developed in humans. Our large prefrontal cortex—a part of the brain with an important role in self-control, self-assessment, and foresight—is a mark of our highly developed capacities in this area.

Although conscience is surely a component of human morality, we are not convinced that this is a "difference in kind" between humans and other animals. Whether or not some correlate of conscience is present in other social mammals, and whether conscience is a prerequisite for moral behavior remain open questions worth deeper consideration. Boehm's work could usefully shape future research into conscience and animal morality: should we look for conscience, and thus morality, in other social mammals that engage in cooperative hunting, such as wolves? Could there have been other evolutionary scenarios in which "systematic and decisive group control" might have served functions other than the distribution of meat?

SPECIES-RELATIVE MORALITY DOES NOT EQUAL "ANYTHING GOES" MORALITY

When people hear us say that morality is species-relative, they may assume that we endorse a philosophical position referred to as moral relativism, which is the view that there are no moral absolutes and that good and bad are nothing more than the conventions of a particular culture or the random preferences of a particular individual. Species-relative morality simply means that we don't look at wolf morality or elephant morality and judge it by some normative standard that we believe holds true for humans. Wolf morality is unique to wolves. We don't judge it at all; we simply describe, observe, and seek to understand. Universally shared patterns of behavior find unique expression in each species, and in each individual.

To help clarify our position, we can borrow a distinction from philosophy between what are called *descriptive* accounts of morality and *normative* accounts. Used descriptively, morality simply refers to a code of conduct put forward by a society to guide the behavior of its members. There is no particular content implied by this definition, no particular behaviors or norms that should be considered right or wrong. What we have done in this book is to give a descriptive account of moral behavior in animals.

On the other hand, our definition of morality does have normative elements. In other words, we say some concrete things about what constitutes moral as opposed to immoral behavior. Moral behavior is other-regarding and prosocial; it is behavior that promotes harmonious co-existence by avoiding harm to others and providing others with help. Norms of behavior that regulate social interactions are found in humans and animals alike. And these norms seem to be universal: in those animal societies in which morality has evolved, we see a common suite of behaviors.

We've argued for what we call species-relative and situational morality (keeping in mind that there are also within species differences in how social norms are understood and expressed). But it doesn't follow from our species-relative account that we think human morality is purely relative, that there are no standards of behavior, no moral truths that may reflect our common aspirations and capacities better than others. For humans, it may not be enough simply to claim that morality is the set of social arrangements that maintain social harmony. Although current social arrangements may indeed allow a certain kind of equilibrium, they may also be unjust for certain segments of the society, or brutalize some segments, or encourage xenophobia. The civil rights movement and women's suffrage both challenged the prevailing societal arrangements. Both movements disrupted social harmony, but we tend to think that the disruption was positive, that our society "evolved" in some important sense from these internal struggles.

An evolutionary approach to morality can help with the problem of relativism, because core behavior patterns are found in all human societies, and they're also found throughout animal societies in nature. These core behavior patterns may be heavily instinctual or hardwired; here, universal norms are likely to emerge, such as an instinctive empathic or altruistic response. Other, more species-relative norms may be peculiar to culture and place. There is room for both universals and for moral innovation.

Although much of the research we've cited in Wild Justice speaks to human moral behavior, we need to be very clear that we're not trying to provide a genealogy of human morality; we don't offer any hypotheses about where human morality comes from or why certain norms seem to persist over time and across cultures. We've made a number of arguments about what morality is and is not like, and have proffered that Western philosophical accounts of morality are outdated in important respects, for example in ascribing too much volition and intentionality to moral behavior. And certainly much of the research on which Wild Justice relies has implications for thinking about human morality, and we have mentioned in various places research on human empathy, altruism, and fairness, as it relates to understanding the behavior of animals. But we want to be absolutely clear that our interest is in animals and in the moral systems that function within animal societies. For those interested in the evolution of human morality, there have been numerous books written about the evolution of human cooperation and human moral behavior (see the first note to this chapter).

ANIMAL RIGHTS AND WRONGS

One of the most obvious questions raised by Wild Justice concerns our ethical responsibilities toward animals. Does ascribing morality to animals mean that our ethical responsibilities to them need to undergo reconsideration?

Scientific data on animal morality do not lead inevitably to a particular conclusion about how we should treat animals or what our relationship to them should be. A scientific description cannot, according to the rules of formal logic, generate a moral imperative. We could easily say, "Animals have morality" and go on treating them just the way we do. Impersonal logic, however, can lead, and has led, to the most egregious treatment of animals in a wide variety of venues.

It is worth noting that modern scientific research on animals, as well as the industrial farming of animals, has traditionally been justified by a scientific description of what animals are like. It's long been asserted as scientific fact that animals don't have complex thoughts or rich emotional lives. It is therefore, the old logic goes, morally acceptable to use animals however we please. As it turns out, the scientific description of the cognitive and emotional capacities of animals has undergone a major sea change in the last decade, and the old logic no longer works. In fact, the new logic imposes strong constraints on how we interact with other animals.

A scientifically accurate description has the power to alter our perception of reality and can thus alter our moral responses. Martin Hoffman, a psychologist who devoted his life to the study of empathy, believes that the empathic predisposition gains maturity and depth, as well as stability and breadth of scope, through veridical perception and deep discernment. In other words, the more cognitively sophisticated our perception of reality, the more deeply and accurately empathic we become. Interestingly, research has suggested that empathic understanding also leads to enhanced critical and moral reasoning. More careful and scientifically accurate description of the lives of animals may lead to increased sensitivity to their needs. If we understand animals to have rich emotional and social lives—to deeply feel many of the same emotions that we do, and to be as connected emotionally with family and friends as we are—it may increase our capacity to empathize with them and feel more "ruth" for the suffering they experience.

SOME CLOSING THOUGHTS ON A YOUNG SCIENCE: MOVING TOWARD A NEW SYNTHESIS

Let's move now toward closing the book on wild justice and at the same time opening up channels for much-needed discussion.

This book crept upon us gradually, a slow and dawning realization that we were really onto something big. We'd both been immersed for different reasons in the literature of ethology, happily working on other projects. Neither of us was particularly looking for moral behavior in animals. But for each of us, quite independently, the data offered too many tantalizing clues to be ignored. At the same time it seemed like a radical step, one that might even endanger our professional credibility, to take the data any further than others had gone. We both felt like that first penguin, stepping out onto thin ice. But we decided that the benefits of doing so greatly outweighed the risks. The amount of interest in questions about morality in animals and the origins of human morality has grown over the past few years, and there is no doubt that interest in wild justice will continue to flourish.

We didn't begin with a definition of morality and then sift through the data to find behavior that fit our description. Rather, we began with a huge pile of descriptive and empirical data about animal behavior, and allowed the data to do the talking, and we tried to let the animals speak for themselves. Only after total immersion did we begin to formulate the hypothesis that some species of animals display a suite of behaviors that, taken together, constitute a system of morality. Certain core behaviors are common across a whole range of species, including humans, and seem to naturally cluster into three general groupings: fairness behaviors, cooperative and altruistic behaviors, and empathic behaviors. Within each of these clusters, and across the whole suite of morality, we see a spectrum of behavioral possibilities, from simple to complex.

Knowledge of animal morality, both its cognitive and affective underpinnings, and social behaviors is relatively young, and continuing work in ethology, animal behavior, and biology will help solve some of the scientific puzzles. At the same time, we are starting to understand a great deal more about the neural basis of human morality, which is shedding light on long-standing philosophical disagreements, such as the role of emotion and cognition in shaping moral behavior. Ethologists, neurobiologists, cognitive psychologists, and other scientists are beginning to collaborate with philosophers and theologians, and are exploring the implications of this new science. All of this together is ushering in a revolution and a new synthesis in how we understand our animal kin and ourselves.

A great deal of work remains to be done to reach a mature understanding of the moral lives of animals, and that's what makes questions about animal morality so exciting. Indeed, this is a project that will likely never reach completion because, as with most research, answers beget more questions. To begin, we need to pay attention to the nitty-gritty details of what animals do in various social venues. We also need individuals to be able to express the full range of behavior patterns they use in their social encounters. In a sense, we need to interview animals and give close attention to the stories they constantly share with us. For sure, other species will also always remain at least a little mysterious. However, much of their behavior is a public affair for all to see, hear, and smell. New research will unlock even more doors and reveal to us new worlds that, for now, are beyond imagination. And likely many questions will unravel somewhere along the line, leaving puzzles with missing pieces, yet with enough of the picture filled in to give us a glimpse of the whole. But for now there can be no doubt that many animals are moral beings. We're not alone in the moral arena. It's just too stingy and incorrect to take this narrow point of view.

WHAT GOES AROUND COMES AROUND

Let's return to where we began. A teenage female elephant nursing an injured leg is knocked over by a rambunctious, hormone-laden teenage male. An older female sees this happen, chases the male away, and goes back to the younger female and touches her sore leg with her trunk. Eleven elephants rescue a group of captive antelope in KwaZula-Natal; the matriarch undoes all of the latches on the gates of the enclosure with her trunk and lets the gate swing open so the antelope can escape. A rat in a cage refuses to push a lever for food when it sees that another rat receives an electric shock as a result. A male diana monkey who has learned to insert a token into a slot to obtain food helps a female who can't get the hang of the trick, inserting the token for her and allowing her to eat the food reward. A female fruit-eating bat helps an unrelated female give birth by showing her how to hang in the proper way. A cat named Libby leads her elderly, deaf, and blind dog friend, Cashew, away from obstacles and to food. In a group of chimpanzees at the Arnhem Zoo in The Netherlands individuals punish other chimpanzees who are late for dinner because no one eats until everyone's present. A large male dog wants to play with a younger and more submissive male. The big male invites his younger partner to play and restrains himself, and biting his younger companion gently and allowing him to bite gently in return. Do these examples show that animals display moral behavior, that they can be compassionate, empathic, altruistic, and fair? Do animals have a kind of moral intelligence? Yes, they do.

ACKNOWLEDGMENTS

We both thank Christie Henry for her patience and commitment to this project. She's been a gem. Dmitri Sandbeck and Pete Beatty also helped in the preparation of our book, Kate Frentzel did a wonderful job of copy editing, and Levi Stahl helped with PR. Over the years, Marc's conversations with Colin Allen, Dale Jamieson, Donald Griffin, Jane Goodall, Susan Townsend, Michael Lemonick, Bruce Gottlieb, David Hatfield, Christine Caldwell, Marjorie Bekoff, Robert Adler, and the men in his classes at the Boulder County Jail helped to shape many of the ideas in this book, but none of these people is to be blamed for what he/we made of them. Jessica thanks the colleagues and friends who have lent an ear and an open mind to *Wild Justice*. In particular, she's grateful for conversations at the ISEE/IAEP Allenspark (Colorado) conference, and thanks especially to Baylor Johnson for long email correspondences about agency and other philosophical troubles. Baylor and an anonymous referee offered extremely helpful comments on an earlier version of this book. Tom Mangelsen (Images of Nature; http://www.mangelsen.com/) kindly provided three photos and we are indebted to his generosity. We're also grateful to Iain Douglas-Hamilton and Shivani Bhalla for providing the photographs of Grace and Eleanor. Thanks also to the rewilding group (Lynne, Margot, Rick, and our occasional visitors) that helped nurture the seeds of wild justice; Roger and Alexandra for listening, questioning, and reading; Benjamin for his priceless practical insight. Finally, thanks to Chris for his steadfast love and Sage for her tender heart.

NOTES

PREFACE

PAGE x *Cornell University historian Dominick LaCapra has claimed that the twenty-first century will be the century of the animal.* This comment was related to Marc by Walter Putnam after a lecture Marc gave at the University of New Mexico, February 6, 2008. Professor LaCapra told Marc, "You may quote me and simply attribute the remark to a comment made after a lecture at the University of New Mexico" (email correspondence, February 9, 2008).

PAGE xi *A cover story in* Time *magazine.* Jeffrey Kluger, "What Makes Us Moral?" *Time,* December 3, 2007, 54–60: http://www.time.com/time/specials/2007/article/0,28804,1685055_1685076_1686619,00.html.

PAGE xii *As we were completing this book.* "Editorial: Survival of the Nicest," *New Scientist,* November 3, 2007; David Sloan Wilson and Edward O. Wilson, "Survival of the Selfless," *New Scientist,* November 3, 2007, 42–46.

CHAPTER 1

PAGE 4 *After about twenty-five years of research.* Jane Goodall. *The Chimpanzees of Gombe: Patterns of Behavior* (Cambridge, MA: Harvard University Press, 1986), 357.

PAGE 10 *He says of this behavior.* B. Heinrich, *Mind of the Raven: Investigations and Adventures with Wolf-Birds* (New York: Cliff Street Books, 1999), 282.

PAGE 15 *"The animal world has its full share."* W. T. Hornaday, *The Minds and Manners of Wild Animals* (New York: Charles Scribner's Sons, 1922), 243.

PAGE 16 *"The guy simply wasn't nice."* R. M. Sapolsky, *A Primate's Memoir* (New York: Touchstone Books, 2002), 234.

PAGE 17 *To wit, in their 2006 review of comparative rates of violence.* Richard W. Wrangham, Michael Wilson, and Martin Muller, "Comparative Rates of Violence in Chimpanzees and Humans," *Primates* 47 (2006): 14–26 (21–22).

PAGE 18 *"Cruelty (from the Latin crudelem, 'morally rough')."* V. Nell, "Cruelty's Rewards: The Gratifications of Perpetrators and Spectators," *Behavioral and Brain Sciences* 29(2006): 211.

CHAPTER 2

PAGE 29 *In 2006, mirror neuron researcher Giacomo Rizzolatti was quoted.* Sandra Blakeslee, "Cells That Read Minds," *New York Times*, January 10, 2006: http://www.nytimes .com/2006/01/10/science/10mirr.html?pagewanted=print. See also V. Gallese, "Mirror Neurons: From Grasping to Language," *Consciousness Bulletin* (Fall 1998): 3–4; V. Gallese, P. F. Ferrari, E. Kohler, and L. Fogassi, "The Eyes, the Hand, and the Mind: Behavioral and Neurophysiological Aspects of Social Cognition," in *The Cognitive Animal*, ed. M. Bekoff, C. Allen, and G. M. Burghardt (Cambridge, MA: MIT Press, 2002), 451–61; V. Gallese and A. Goldman, "Mirror Neurons and the Simulation Theory of Mind-Reading," *Trends in Cognitive Science* 2 (1998):493–501.

PAGE 29 *Neuroscientist V. S. Ramachadran claims.* Mirror neurons and imitation learning as the driving force behind "the great leap forward" in human evolution: http:// www.edge.org/3rd_culture/ramachandran06/ramachandran06_index.html.

PAGE 30 *To sum up the significance of spindle cells.* http://lifeboat.com/ex/bios.lori .marino.

PAGE 34 *"Without emotion you have a dead study."* George Schaller, "Feral and Free— An Interview with George Schaller," *New Scientist*, April 5, 2007, 46–47.

PAGE 34 *Consider, too, the reflections of George Schaller.* George Schaller, "Feral and Free—An Interview with George Schaller," *New Scientist*, April 5, 2007, 46–47.

PAGE 39 *Reed Montague notes that the caudate nucleus.* Greg Miller, "Economic Game Shows How the Brain Builds Trust," *Science* 308 (no. 5718): 36. http://www .sciencemag.org/cgi/content/summary/308/5718/36a.

PAGE 39 *Kagan notes that, "there is no large body of impeccable, interrelated facts."* J. Kagan, *Three Seductive Ideas* (Cambridge, MA: Harvard University Press, 1998): 11.

PAGE 41 *Consider the words of Sarita Siegel.* S. Siegel, "Reflections on Anthropomorphism in the Enchanted Forest," in *Thinking with Animals: New Perspectives on Anthropomorphism*, ed. L. Daston and G. Mitman (New York: Columbia University Press, 2005), 221.

PAGE 41 *Canadian biologist Hal Whitehead.* H. Whitehead, *Sperm Whales: Social Evolution in the Oceans* (Chicago: University of Chicago Press, 2004), 370–71.

PAGE 42 *"Yes, we are human and cannot avoid the language."* S. J. Gould, "A Lover's Quarrel," in *The Smile of a Dolphin: Remarkable Accounts of Animal Emotions*, ed. M. Bekoff (New York: Random House/Discovery Books, 2000), 17.

PAGE 42 *In an interview in Salon magazine.* Douglas Cruickshank, "Robert Sapolsky," http://dir.salon.com/story/people/conv/2001/05/14/sapolsky/index.html?pn=2.

PAGE 47 *Sociality, then, refers to "the compromises that individuals make."* R. Sussman and A. R. Chapman, eds., *The Origins and Nature of Sociality* (Chicago: Walter de Gruyter, Inc.), 10.

PAGE 48 *For example, Robert Sapolsky studied.* Mark Shwartz, *Stanford Report*, March 7, 2007, http://news-service.stanford.edu/news/2007/march7/sapolskysr-030707.

PAGE 48 *Animals have various means of maintaining social order.* F. de Waal, *Good-Natured The Origins of Right and Wrong in Humans and Other Animals* (Cambridge, MA: Harvard University Press, 1996), 207.

PAGE 49 *Jerome Kagan writes,: The defenders."* J. Kagan, *Three Seductive Ideas* (Cambridge, MA: Harvard University Press, 1998), 52.

PAGE 53 *As the research of.* P. Leyhausen, *Cat Behavior* (New York: Garland, 1978).

CHAPTER 3

PAGE 56 *He wrote, "They probably evolved."* Frans de Waal, "How Animals Do Business," *Scientific American* 292: 74, 76.

PAGE 57 *In Mutual Aid Kropotkin laments.* Peter Kropotkin, *Mutual Aid: A Factor of Evolution* (1902; reprinted 2006 by BiblioBazaar), 22.

PAGE 57 *Sussman and his colleagues concluded.* Robert W. Sussman, Paul A. Garber, and James M. Cheverud, "Importance of Cooperation and Affiliation in the Evolution of Primate Sociality," *American Journal of Physical Anthropology* 128 (2005): 92.

PAGE 59 *"Cooperation," Nowak says, "is the secret."* Martin A. Nowak, "Five Rules for the Evolution of Cooperation," *Science* 314 (2006): 1560–63.

PAGE 59 *De Waal argues for a similar point of emphasis.* Frans B. M. de Waal, "Morality and the Social Instincts: Continuity with the Other Primates" (Tanner Lectures on Human Values, Princeton University, November 19–20, 2003), 13.

PAGE 60 *As philosopher Elliott Sober and evolutionary biology David Sloan Wilson.* E. Sober and D. S. Wilson, *Unto Others: The Evolution and Psychology of Unselfish Behavior* (Cambridge, MA: Harvard University Press, 1998), 17.

PAGE 63 *Writing about humans, Taylor observes.* Shelley Taylor, *The Tending Instinct: How Nurturing is Essential for Who We Are and How We Live* (New York: Henry Holt and Company, 2002), 13.

PAGE 66 *We believe, along with other biologists such as David Sloan Wilson and Edward O. Wilson.* See, for example, Nicholas Wade, "Taking a Cue from Ants on the Evolution of Humans," *New York Times*, July 15, 2008: http://www.nytimes.com/2008/07/15/science/15wils.html?_r=1&scp=1&sq=wade%20ants&st=cse&oref=slogin.

PAGE 77 *Neurobiologist Jaak Panksepp has suggested.* In Taylor, *The Tending Instinct*, 82.

PAGE 80 *"Either the cognitive implications attributed to primates."* Christine M. Drea and Laurence G. Frank, "The Social Complexity of Spotted Hyenas," in *Animal Social Complexity*, ed. F. B. M. de Waal and P. L. Tyack (Cambridge, MA: Harvard University Press, 2003), 121–48.

PAGE 81 *Brian Hare's work, for example.* "Social Tolerance Allows Bonobos to Outperform Chimpanzees on a Cooperative Task," *Science Daily*, March 9, 2007: http://www.sciencedaily.com/releases/2007/03/070308121928.htm.

PAGE 82 R. E. Hudson, J. E. Aukema, C. Rispe, and D. Roze, "Altruism, cheating, and anticheater adaptations in cellular slime molds," *American Naturalist* 160 (2002), 31.

CHAPTER 4

PAGE 86 *Frans de Waal said of Langford's research.* In Benedict Carey, "Message from Mouse to Mouse: I Feel Your Pain," *New York Times*, July 4, 2006.

PAGE 86 *Jaak Panksepp remarked, "If it turns out."* In Ishani Ganguli, "Mice Show Evidence of Empathy," *Scientist*, June 30, 2006: http://www.the-scientist.com/news/display/23764/.

PAGE 88 *As Preston and de Waal explain.* Stephanie D. Preston and Frans B. M. de Waal, "Empathy: Its Ultimate and Proximate Bases," *Behavioral and Brain Sciences* 25 (2002) 1–72.

PAGE 92 *Roger Highfield, writing in the UK Telegraph.* Roger Highfield, "Orangutans Share a Joke Too," *Telegraph*, December 12, 2007: http://www.telegraph.co.uk/earth/main.jhtml?xml=/earth/2007/12/12/scioran112.xml.

PAGE 95 *Darwin tells a number of stories.* Charles Darwin, *The Descent of Man, and Selection in Relation to Sex* (New York: Penguin Classics, 1871/2004), 126.

PAGE 95 *He concludes, "Any animal whatever."* Charles Darwin, *The Descent of Man*, 71–72.

PAGE 98 *In an interview, anthropologist Barbara J. King.* Barbara King, *Primatology.net*, January 31, 2007: http://primatology.net/2007/01/31/on-god-gorillas-and-the-evolution-of-religion/.

PAGE 103 *As Douglas-Hamilton writes in his field observation.* Iain Douglas-Hamilton, S. Bhalla, G. Wittemyer, and F. Vollrath, "Behavioural Reactions of Elephants towards a Dying and Deceased Matriarch," *Applied Animal Behaviour Science* 100 (2006): 87–102.

PAGE 105 *After her black rhino companion was shot.* M. Ryan and P. Thornycraft, "Jumbos Mourn Black Rhino Killed by Poachers," *Sunday Independent*, November 18, 2007: http://www.sundayindependent.co.za/.

PAGE 105 *Bradshaw and UCLA neuroscientist Allan Schore report.* Gay A. Bradshaw and Allan N. Schore, "How Elephants are Opening Doors: Developmental Neuroethology, Attachment, and Social Context," *Ethology* 113 (2007): 426–36.

PAGE 109 *"If I pretend to be crying."* Nadia Ladygina-Kohts: http://press.princeton.edu/chapters/s8240.html.

CHAPTER 5

PAGE 110 *"Fairness is only human."* Denise Gellene, "Fairness Is Only Human, Scientists Find," *Los Angeles Times*, October 5, 2007.

PAGE 111 *Boysen notes, "Deviations from their code of conduct."* In Gellene, "Fairness Is Only Human, Scientists Find."

PAGE 111 *So too does recent research by Friederike Range.* Friederike Range, "Effort and Reward: Inequity Aversion in Domestic Dogs?" (Canine Science Forum, Budapest, Hungary, July 2008).

PAGE 112 *"Some wolves are fair."* Robert C. Solomon, *A Passion for Justice* (Lanham, MD: Rowman & Littlefield, 1995), 141.

PAGE 114 *Kiley Hamlin, who along with her colleagues.* Helen Briggs, "Babies 'Show Social Intelligence,'" BBC News, November 21, 2007: http://news.bbc.co.uk/2/hi/science/nature/7103804.stm. See also http://www.nature.com/news/2007/071121/full/news.2007.278.html.

PAGE 125 *In his book* The Descent of Man. Darwin, *The Descent of Man*, 69.

PAGE 125 *Ethologist Jonathan Balcombe says that pleasure.* Jonathan Balcombe, *Pleasurable Kingdom: Animals and the Nature of Feeling Good* (New York: Macmillan, 2006), 9, 11.

PAGE 126 *In his book* Darwin's Cathedral. D. S. Wilson, *Darwin's Cathedral: Evolution, Religion, and the Nature of Society* (Chicago: University of Chicago Press, 2002), 212.

PAGE 127 *Brosnan and de Waal note, "Females pay closer attention" and subsequent quotes.* Sarah F. Brosnan and Frans B. de Waal, "Monkeys Reject Unequal Pay," *Nature* 425 (2003): 297–99.

PAGE 128 *As de Waal notes in* Good Natured. F. B. M. de Waal, *Good Natured* (Cambridge, MA: Harvard University Press, 1996), 159.

PAGE 128 *In a 2005* Scientific American *essay Frans de Waal argued.* Frans B. M. de Waal, "How Animals Do Business," *Scientific American* 292 (2005): 73–79.

PAGE 128 *De Waal clearly recognizes the implications.* F. B. M. de Waal, *Good Natured* (Cambridge, MA: Harvard University Press), 161.

PAGE 131 *To quote Mitchell, a "society of self-interest."* Lawrence E. Mitchell, *Stacked Deck: A Story of Selfishness in America* (Philadelphia: Temple University Press, 1998), 205.

PAGE 131 *"We can start by changing our minds."* Mitchell, *Stacked Deck: A Story of Selfishness in America* (Philadelphia: Temple University Press, 1998), 210.

PAGE 132 *Robert Solomon writes, in* A Passion for Justice. Robert C. Solomon, *A Passion for Justice* (Lanham, MD: Rowman & Littlefield, 1995), 102.

PAGE 132 *He writes, "Being able to care."* Paul Shapiro, "Moral Agency in Other Animals," *Theoretical Medicine* 27 (2006): 357–73.

PAGE 132 *But, he says, "to get to morality."* Frans de Waal, *Believer* interview, September 2007: http://www.believermag.com/issues/200709/?read=interview_dewaal.

PAGE 132 *As the renowned neuroscientist Antonio Damasio has argued.* A. Damasio, *Looking for Spinoza: Joy, Sorrow, and the Feeling Brain* (New York: Harcourt, 2003), 103.

CHAPTER 6

For readers with an interest in the evolution of human morality, we recommend the following books: Robert Axelrod's *The Complexity of Cooperation*, *Moral Sentiments and Material Interests*, edited by Herbert Gintis and his colleagues; *Foundations of Human Sociality*, edited by Joseph Henrich; Robert Hinde's *Why Good Is Good*; Eibl-Eibesfelt's *Human Ethology*; Richard Alexander's *The Biology of Moral Systems*; Elliott Sober and David Sloan Wilson's *Unto Others*; Lee Alan Dugatkin's *The Altruism Equation*; Richard Wright's *The Moral Animal*; Frans de Waal's *Our Inner Ape*; Marc

Hauser's *Moral Minds*; Michael Shermer's *The Science of Good and Evil*; Phillip Clayton and Jeffrey Schloss's *Evolution and Ethics*; and, for a more philosophical take, Frederick Nietzsche's *A Genealogy of Morals* and Shaun Nichol's *Sentimental Rules*.

PAGE 136 *Kunz concluded that "such cooperative behaviour."* New Scientist, June 16, 1994: http://www.newscientist.com/article/mg14219302.900-science-bat-mothers-share-the-birth-experience-.html.

PAGE 139 *"[T]he capacity for normative self-government."* C. Korsgaard, in *Primates and Philosophers*, by F. B. M. de Waal (Princeton: Princeton University Press, 2006), 116.

PAGE 140 *Deacon notes, "[T]he first use of symbolic reference."* T. W. Deacon, *The Symbolic Species: The Co-Evolution of Language and Brain* (New York: W. W. Norton & Company), 322.

PAGE 140 *Although the prominent biologist E. O. Wilson famously argued.* E. O. Wilson, *Sociobiology: The New Synthesis* (Cambridge, MA: Belknap, 1975), 562.

PAGE 144 *Paul Shapiro correctly notes.* Paul Shapiro, "Moral Agency in Other Animals," *Theoretical Medicine* 27 (2006): 357–73.

PAGE 145 *In The Descent of Man, Darwin wrote.* Darwin, *The Descent of Man*, 127.

PAGE 146 *"The one course ought to have been followed."* Darwin, *The Descent of Man*, 123.

PAGE 146 *"Moral judgments," he says, "depend on comparisons."* R. Hinde *Why Good Is Good*, 53.

PAGE 146 *As Boehm says, "bands had to gang up."* C. H. Boehm, "Conscience Origins, Sanctioning Selection, and the Evolution of Altruism in *Homo sapiens*" (submitted manuscript, personal communication).

GENERAL REFERENCES

This list contains both references that are included in the text and others that are not, but that are relevant to our discussion of wild justice.

Adolphs, Ralph. 1999. Social cognition and the human brain. *Trends in Cognitive Sciences* 3:469–79.

Alexander, Richard D. 1987. *The Biology of Moral Systems*. New York: Aldine de Gruyter.

Allen, C. 2001. Cognitive relatives and moral relations. In *Great Apes and Humans at an Ethical Frontier*, ed. Beck, B. B., T. S. Stoinski, M. Hutchins, T. S. Maple, B. Norton, A. Rowan, B. F. Stevens, and A. Arluke. Washington, D.C.: Smithsonian Institution Press.

Allen, C., and M. Bekoff. 1997. *Species of Mind*. Cambridge, MA: MIT Press.

———. 2005. Animal play and the evolution of morality: An ethological approach. *Topoi* 24:125–35.

Appiah, K. A. 2008. *Experiments in Ethics*. Cambridge, MA: Harvard University Press.

Aureli, F., ed.. 2000. *Natural Conflict Resolution*. Berkeley: University of California Press.

Axelrod, Robert. 2006. *The Evolution of Cooperation*. Rev. ed. New York: Perseus Books Group.

Axelrod, Robert, and William Hamilton. 1981. The evolution of cooperation. *Science* 211:1390–96.

Balcombe, Jonathan. 2006. *Pleasurable Kingdom: Animals and the Nature of Feeling Good*. New York: Macmillan.

Balcombe, Jonathan P., Neal D. Barnard, and Chad Sandusky. 2004. Laboratory routines cause animal stress. *Contemporary Topics*, American Association for Laboratory Science 43:42–51.

Baldwin, Ann and Marc Bekoff. 2007. Too stressed to work. *New Scientist*, June 2:24.

Barrett, L., S. P. Henzi, T. Weingrill, J. E. Lycett, and R. A. Hill. 1999. Market forces predict grooming reciprocity in female baboons. *Proceedings of the Royal Society of London* 266:665–70.

Bates, L. A., and R. W. Byrne. 2007. Creative or created: Using anecdotes to investigate animal cognition. *Methods* 42:12–21. http://www.st-andrews.ac.uk/ffiwww_sp/people/personal/rwb/publications/2007%20Bates%20Byrne%20Methods.pdf.

Bateson, Patrick. 2000. The biological evolution of cooperation and trust. In *Trust: Making and Breaking Cooperative Relations*, ed. Diego Gambetta, 14–30. Oxford: Department of Sociology, University of Oxford. http://www.sociology.ox.ac.uk/papers/bateson14–30.pdf.

Batson, C. Daniel. 1991. *The Altruism Question: Toward a Social-Psychological Answer.* Hillsdale, NJ: Lawrence Erlbaum Associates.

Bearzi, M., and C. B. Stanford. 2008. *Beautiful Minds: The Parallel Lives of Great Apes and Dolphins.* Cambridge, MA: Harvard University Press.

Bekoff, M. 1996. Cognitive ethology, vigilance, information gathering, and representation: Who might know what and why? *Behavioural Processes* 35:225–37.

———. 2005. Vigilance, flock size, and flock geometry: Information gathering by Western Evening Grosbeaks (Aves, fringillidae), *Ethology* 99:150–61.

———. 2007. *The Emotional Lives of Animals.* Novato, CA: New World Library.

Bekoff, M., C. Allen, and G. M. Burghardt, eds. 2002. *The Cognitive Animal: Empirical and Theoretical Perspectives on Animal Cognition.* Cambridge, MA: MIT Press.

Bekoff, M., and John A. Byers, eds. 1998. *Animal Play: Evolutionary, Comparative, and Ecological Perspectives.* Cambridge: Cambridge University Press.

Bekoff, M., and M. C. Wells. 1986. Social behavior and ecology of coyotes. *Advances in the Study of Behavior* 16:251–338.

Blum, Deborah. 2004. *Love at Goon Park: Harry Harlow and the Science of Affection.* Cambridge, MA: Perseus Publishing.

Boehm, Christopher. 1999. *Hierarchy in the Forest: The Evolution of Egalitarian Behavior.* Cambridge, MA: Harvard University Press.

———. Conscience origins, sanctioning selection, and the evolution of altruism in *Homo sapiens* (submitted manuscript, personal communication).

Borba, M. 2001. *Building Moral Intelligence: The Seven Essential Virtues That Teach Kids to Do the Right Thing.* San Francisco: Jossey-Bass.

Bradshaw, G., A. N. Schore, J. L. Brown, J. H. Poole, and C. Moss. 2005. Elephant breakdown. *Nature* 433:807.

Bradshaw, G. A., and A. N. Schore. 2007. How elephants are opening doors: Developmental neuroethology, attachment, and social context. *Ethology* 133:426–36.

Brosnan, S. F. 2006. Nonhuman species' reactions to inequity and their implications for fairness. *Social Justice Research* 19:153–85.

Brosnan, S. F., and F. B. M. de Waal. 2002. A proximate perspective on reciprocal altruism. *Human Nature* 13:129–52.

Brosnan, S. F., and Frans B. de Waal. 2003. Monkeys reject unequal pay. *Nature* 425: 297–99.

Brosnan, S. F., H. Schiff, and F. B. de Waal. 2004. Tolerance for inequity may increase with social closeness in chimpanzees. *Proceedings of the Royal Society B.* 1560: 253–58.

Bshary, R., and A. S. Grutter. 2006. Image scoring and cooperation in a cleaner fish mutualism. *Nature* 441:975–78.

Bshary, R., A. Hohner, K. Ait-el-Djoudi, and H. Fricke. 2006. Interspecific communicative and coordinated hunting between groupers and giant Moray eels in the Red Sea. *PLoS Biology* 4 (12): e431.

Burgdorf, J., and J. Panksepp. 2001. Tickling induces reward in adolescent rats. *Physiology and Behavior* 72(1–2): 167–73.

Burghardt, G. M. 2005. *The Genesis of Animal Play: Testing the Limits*. Cambridge, MA: MIT Press.

Byrne, R. W. 1994. The evolution of intelligence. In *Behaviour and Evolution*, ed. P. J. B. Slater and T. R. Halliday, 223–65. Cambridge: Cambridge University Press.

Byrne, R. W., and N. Corp. 2004. Neocortex size predicts deception rate in primates. *Proceedings of the Royal Society B* 271:1693–99.

Byrne, R. W., and A. Whiten, eds. 1988. *Machiavellian Intelligence: Social Expertise and the Evolution of Intellect in Monkeys, Apes, and Humans*. Oxford: Clarendon Press.

Cheney, D. L., and R. M. Seyfarth. 1990. *How Monkeys See the World*. Chicago: University of Chicago Press.

———. 2007. *Baboon Metaphysics: The Evolution of a Social Mind*. Chicago: University of Chicago Press.

Church, R. 1959. Emotional reactions of rats to the pain of others. *Journal of Comparative and Physiological Psychology* 52:132–34.

Clayton, P. and J. Schloss, eds. 2004. *Evolution and Ethics: Human Morality in Biological and Religious Perspective*. Grand Rapids: William B. Eerdmans.

Clutton-Brock, T. H., and Paul H. Harvey. 1980. Primates, brains, and ecology. *Journal of the Zoological Society of London* 190:309–23.

Clutton-Brock, T. H., and G. A. Parker. 1995. Punishment in animal societies. *Nature* 373:209–16.

Coetzee, J. M. 1999. *The Lives of Animals*. Princeton: Princeton University Press.

Cools, A., A. van Hout, and M. Nelissen. 2008. Canine reconciliation and third-party-initiated postconflict affiliation: Do peacemaking social mechanisms in dogs rival those of higher primates? *Ethology* 114:53–62.

Costa, J. T. 2006. *The Other Insect Societies*. Cambridge, MA: Belknap.

Creager, A. N. H., and W. Chester Jordan, eds. 2002. *The Animal/Human Boundary: Historical Perspectives*. Rochester: University of Rochester Press.

Damasio, A. 1994 . *Descartes' Error: Emotion, Reason, and the Human Brain*. New York: Penguin.

———. 1999. *The Feeling of What Happens: Body and Emotion in the Making of Consciousness*. New York: Harcourt Brace.

———. 2003. *Looking for Spinoza: Joy, Sorrow, and the Feeling Brain*. New York: Harcourt.

Datson, L., and G. Mitman. 2005. *Thinking with Animals: New Perspectives on Anthropomorphism*. New York: Columbia University Press.

Davidson, R. J., K. R. Scherer, and H. Hill Goldsmith, eds. 2003. *Handbook of Affective Sciences*. New York: Oxford University Press.

Dawkins, R. 1976. *The Selfish Gene*. New York: Oxford University Press.

Deacon, T. W. 1997. *The Symbolic Species: The Co-Evolution of Language and Brain*. New York: W. W. Norton & Company.

Decety, J., P. Jackson, J. Sommerville, T. Chaminade, and A. Meltzoff, 2004. The neural bases of cooperation and competition: an fMRI investigation. *Neuroimage* 23:744–51.

Decety, J., and P. L. Jackson. 2004. The functional architecture of human empathy. *Behavioral and Cognitive Neuroscience Reviews* 3:71–100.

Decety, J. P., and Philip Jackson. 2006. A social-neuroscience perspective on empathy. *Current Directions in Psychological Science* 15:54–58.

de Quervain, D., U. Fischbacher, V. Treyer, M. Schellhammer, U. Schnyder, A. Buck, and E. Fehr. 2004. The neural basis of altruistic punishment. *Science* 305:1254–58.

de Vignemont, F., and T. Singer. 2006. The empathic brain: How, when and why? *Trends in Cognitive Sciences* 10:435–41.

de Waal, F. B. M. 1982. *Chimpanzee Politics: Power and Sex among Apes*. Baltimore: Johns Hopkins University Press.

———. 1996. *Good Natured: The Origins of Right and Wrong in Humans and Other Animals*. Cambridge, MA: Harvard University Press.

———. 2001. Do humans alone "feel your pain"? *The Chronicle of Higher Education*, October 26.

———. 2005a. *Our Inner Ape: A Leading Primatologist Explains Why We Are Who We Are*. New York: Riverhead.

———. 2005b. How Animals Do Business. *Scientific American* 292 (4): 73–79.

———. 2006. *Primates and Philosophers*. Princeton: Princeton University Press.

de Waal, F. B. M., and J. J. Pokorny. 2005. Primate conflict resolution and its relation to human forgiveness. In *Handbook of Forgiveness*, ed. E. L. Worthington, Jr., 17–32. New York: Brunner-Routledge.

de Waal, F. B. M., and P. L. Tyack, eds. 2003. *Animal Social Complexity: Intelligence, Culture, and Individualized Societies*. Cambridge, MA: Harvard University Press.

Doris, J. M., and S. P. Stich. 2005. As a matter of fact: Empirical perspectives on ethics. In *The Oxford Handbook of Contemporary Analytic Philosophy*, ed. F. Jackson and M. Smith, 114–52. New York: Oxford University Press. http://www.rci.rutgers.edu/ffistich/Publications/Papers/05-Jackson-Chap-05.pdf.

Douglas-Hamilton, I., S. Bhalla, G. Wittemyer, and F. Vollrath. 2006. Behavioural reactions of elephants towards a dying and deceased matriarch. *Applied Animal Behaviour Science* 100:87–102.

Drea, C. M., and L. G. Frank. 2003. The social complexity of spotted hyenas. In *Animal Social Complexity*, ed. F. B. M. de Waal and P. L. Tyack, 121–48. Cambridge, MA: Harvard University Press.

Dudzinski, Kathleen and Toni Frohoff. 2008. *Dolphin Mysteries:Unlocking the Secrets of Communication*. New Haven, CT: Yale University Press.

Dugatkin, L. A. 1999. *Cheating Monkeys and Citizen Bees: The Nature of Cooperation in Animals and Humans*. New York: The Free Press.

————. 2006a. Trust in fish. *Nature* 441:937–38.

————. 2006b. *The Altruism Equation: Seven Scientists Search for the Origins of Goodness.* Princeton: Princeton University Press.

Dugatkin, L. A., and M. S. Alfieri. 2002. A cognitive approach to the study of animal cooperation. In *The Cognitive Animal,* ed. M. Bekoff, C. Allen, and G. M. Burghardt, 413–19. Cambridge, MA: MIT Press.

Dugatkin, L. A., and M. Bekoff. 2003. Play and the evolution of fairness: A game theory model. *Behavioural Processes* 60:209–14.

Dunbar, R. 1998. *Grooming, Gossip, and the Evolution of Language.* Cambridge, MA: Harvard University Press.

Ehrlich, P. 2002. *Human Natures: Genes, Cultures, and the Human Prospect.* New York: Penguin.

Eisenberg, N. 1986. *Altruistic emotion, cognition, and behavior.* Hillsdale, NJ: Lawrence Erlbaum.

Emery, N., and N. S. Clayton. 2004. The mentality of crows: Convergent evolution of intelligence in corvids and apes. *Science* 306:1903–7.

Evans, E. P. 1906. *The Criminal Prosecution and Capital Punishment of Animals.* New York: E. P. Dutton.

Fagen, R. M. 1981. *Animal Play Behavior.* New York: Oxford University Press.

Fehr, E., and A. Damasio. 2005. On brain trust. *Nature* 435:571–72.

Fehr, E., and S. Gächter. 2000. Fairness and retaliation: The economics of reciprocity. *Journal of Economic Perspectives* 14:159–81.

Fiske, A. P. 1992. The four elementary forms of sociality: Framework for a unified theory of social relations. *Psychological Review* 99:689–723.

Flack, J. C., and F. B. M. de Waal. 2000. "Any animal whatever": Darwinian building blocks of morality in monkeys and apes. *Journal of Consciousness Studies* 7:1–29.

Fox, M. W. 1969. A comparative study of the development of facial expressions in canids: Wolf, coyote and foxes. *Behaviour* 36:49–73.

Frank, S. A. 1998. *Foundations of Social Evolution.* Princeton: Princeton University Press.

Fraser, O. N., D. Stahl, and F. Aureli. 2008. Stress reduction through consolation in chimpanzees. *Proceedings of the National Academic of Sciences* 105:8557–62.

Gardner, A., and S. A. West. 2004. Spite among siblings. *Science* 305:1413–14.

Gardner, H. 1996. *Multiple Intelligences.* Cambridge, MA: Perseus.

Gazzaniga, M. 1992. *Nature's Mind: The Biological Roots of Thinking, Emotions, Sexuality, Language, and Intelligence.* New York; Penguin.

————. 2005. *The Ethical Brain.* New York: Dana Press.

Gellene, D. 2007. Fairness is only human, scientists find. *Los Angeles Times,* October 5.

Gervais, Matthew, and David Sloan Wilson. 2005. The evolution and functions of laughter and humor: A synthetic approach. *Quarterly Review of Biology* 80: 395–430.

Ghiselin, M. T. 1997. *Metaphysics and the Origin of Species.* Albany: SUNY Press.

Gibbs, J. C. 2003. *Moral Development and Reality: Beyond the Theories of Kohlberg and Hoffman.* Thousand Oaks, CA: Sage Publications.

Gintis, H., S. Bowles, R. Boyd, and E. Fehr. 2005. *Moral Sentiments and Material Interests: The Foundations of Cooperation in Economic Life.* Cambridge, MA: MIT Press.

Goleman, D. 1995. *Emotional Intelligence.* New York: Bantam Books.

———. 2006. *Social Intelligence: The New Science of Human Relationships.* New York: Bantam Books.

Goodall, J. 1986. *The Chimpanzees of Gombe: Patterns of Behavior.* Cambridge, MA: Harvard University Press.

Gray, H. M., K. Gray, and D. M. Wegner. 2007. Dimensions of mind perception. *Science* 315:619.

Greene, J., and J. Haidt. 2002. How (and where) does moral judgment work? *Trends in Cognitive Sciences* 6:517–23.

Griffin. D. R. 1976/1981. *The Question of Animal Awareness: Evolutionary Continuity of Mental Experience.* New York: Rockefeller University Press.

———. 1992. *Animal Minds.* Chicago: University of Chicago Press.

Haidt, J. 2007. The new synthesis in moral psychology. *Science* 316:998–1002.

Hamilton, W. D. 1964. The genetical evolution of social behaviour I and II. *Journal of Theoretical Biology* 7:1–16 and 7:17–52.

Hammerstein, P., ed. 2003. *Genetic and Cultural Evolution of Cooperation.* Cambridge, MA: MIT Press.

Hansson, M. G. 2008. *The Private Sphere: An Emotional Territory and Its Agent.* New York: Springer.

Harcourt, A. H., and Frans B. M. de Waal, eds. 1992. *Coalitions and Alliances in Humans and other Animals.* Oxford: Oxford University Press.

Hare, B., M. Brown, C. Williamson, and M. Tomasello. 2002. The domestication of social cognition in dogs. *Science* 298:1634–36.

Harlow, H. F. 1958. The nature of love. *American Psychologist* 13:673–85.

Hart, B. L., and L. A. Hart. 1992. Reciprocal allogrooming in impala, *Aepyceros melampus. Animal Behaviour* 44:1073–83.

Hatfield, E., J. T. Cacioppo, and R. L. Rapson. 1994. *Emotional Contagion.* Cambridge: Cambridge University Press.

Hauser, M. D. 2000. *Wild Minds.* New York: Henry Holt and Company.

———. 2006. *Moral Minds: How Nature Designed Our Universal Sense of Right and Wrong.* New York: Harper Collins.

Heinrich, B. 1999. *Mind of the Raven: Investigations and Adventures with Wolf-Birds.* New York: Cliff Street Books.

Heinsohn, R., and C. Packer. 1995. Complex cooperative strategies in group-territorial African lions. *Science* 269:1260–62.

Henrich, J., R. Boyd, S. Bowles, C. Camerer, E. Fehr, and H. Gintis. 2004. *Foundations of Human Sociality: Economic Experiments and Ethnographic Evidence from Fifteen Small-Scale Societies.* New York: Oxford.

Henzi, S. P., and L. Barret. 2002. Infants as a commodity in a baboon market. *Animal Behaviour* 63:915–21.

Hinde, R. A. 1974. *Biological Bases of Human Social Behavior.* New York: McGraw-Hill.

———. 1987. *Individuals, Relationships, and Culture: Links between Biology and the Social Sciences.* Cambridge: Cambridge University Press.

———. 2002. *Why Good Is Good: The Sources of Morality.* New York: Routledge.

Hof, P., and E. van der Gucht. 2006. Whales boast the brain cells that 'make us human.' *New Scientist,* November 27. http://www.newscientist.com/channel/life/dn10661-whales-boast-the-brain-cells-that-make-us-human.html.

Hoffman, Martin. 2000. *Empathy and Moral Development: Implications for Caring and Justice.* Cambridge, MA: Harvard University Press.

Holekamp, K. E. 2006. Questioning the social intelligence hypothesis. *Trends in Cognitive Science* 11:65–69.

Hornaday, W. T. 1922. *The Minds and Manners of Wild Animals.* New York: Charles Scribner's Sons.

Horowitz, A. C. 2002. The behaviors of theories of mind, and a case study of dogs at play. Ph.D. diss., University of California, San Diego.

Horowitz, A. C., and M. Bekoff. 2007. Naturalizing anthropomorphism: Behavioral prompts to our humanizing of animals. *Anthrozoös* 20:23–36.

Hull, R. B. 2006. *Infinite Nature.* Chicago: University of Chicago Press.

Humphrey, N. 1988. The social function of intellect. In Byrne and Whiten 1988, 13–26.

———. 1997. Varieties of altruism and the common ground between them. *Social Research* 64:199–209.

———. 2003. *The Inner Eye: Social Intelligence in Evolution.* New York: Oxford University Press.

Iwaniuk, A., S. M. Pellis, and J. E. Nelson. 2001. Do big-brained animals play more? Comparative analyses of play and relative brain size in mammals. *Journal of Comparative Psychology* 115:29–41.

Jablonka, E., and M. J. Lamb. 2005. *Evolution in Four Dimensions: Genetic, Epigenetic, Behavioral, and Symbolic Variation in the History of Life.* Cambridge, MA: Bradford Books.

Jensen, K., J. Call, and M. Tomasello. 2007a. Chimpanzees are rational maximizers in an ultimatum game. *Science* 318:107–9.

———. 2007b. Chimpanzees are vengeful but not spiteful. *Proceedings of the National Academy of Sciences* 104:13046–51.

Johnson, D., P. Stopka, and D. McDonald. 1999. Ideal flea constraints on group living: Unwanted public goods and the emergence of cooperation. *Behavioral Ecology* 15:181–86.

Jolly, A. 1966. Lemur social behavior and primate intelligence. *Science* 153:501–6.

Joyce, R. 2006. *The Evolution of Morality.* Cambridge, MA: MIT Press.

Kagan, J. 1998. *Three Seductive Ideas.* Cambridge, MA: Harvard University Press.

Kagan, J., and S. Lamb. 1987. *The Emergence of Morality in Young Children.* Chicago: University of Chicago Press.

Katz, L. D., ed. 2000. *Evolutionary Origins of Morality: Cross Disciplinary Perspectives.* Bowling Green, OH: Imprint Academics.

Kelly, D., S. Stich, K. J. Haley, S. J. Eng, and D. M. T. Fessler. 2007. Harm, affect, and the moral/conventional distinction. *Mind and Language* 22:117–31.

Kitchen, Dawn M., and Craig Packer. 1999. Complexity in vertebrate societies. In *Levels of Selection in Evolution*, ed. L. Keller, 176–96. Princeton: Princeton University Press.

Koenigs, M., L. Young, R. Adolphs, D. Tranel, F. Cushman, M. Hauser, and A. Damasio. 2007. Damage to the prefrontal cortex increases utilitarian moral judgments. *Nature* 446:908–11.

Kosfeld, M., M. Heinrichs, P. J. Zak, U. Fischbacher, and E. Fehr. 2005. Oxytocin increases trust in humans. *Nature* 435:673–76.

Kropotkin, P. 1902/2006. *Mutual Aid: A Factor of Evolution*. Repr. BiblioBazaar.

Kunz, T. H., A. L. Allgaier, J. Seyjagat, and R. Caliguiri. 1994. Allomaternal care: Helper-assisted birth in the Rodrigues fruit bat, *Pteropus rodricensis* (Chiroptera: Pteropodidae). *Journal of Zoology* 232:691–700.

Langford, D. J., S. E. Crager, Z. Shehzad, S. B. Smith, S. G. Sotocinal, J. S. Levenstadt, M. L. Chanda, D. J. Levitin, and J. S. Mogil. 2006. Social modulation of pain as evidence for empathy in mice. *Science* 312:1967–70.

Lewis, K. P. 2000. A comparative study of primate play behaviour: Implications for the study of cognition. *Folia Primatologica* 71:417–21.

Lewis, M., and J. M. Haviland-Jones. 2000. *Handbook of Emotions*. 2nd ed. New York: The Guilford Press.

Lewis, R. 2002. Beyond dominance: The importance of leverage. *Quarterly Review of Biology* 77:149–64.

Leyhausen, P. 1978. *Cat Behavior*. New York: Garland.

Libet, B. 2004. *Mind Time: The Temporal Factor in Consciousness*. Cambridge, MA: Harvard University Press.

Lyons, D. E., L. R. Santos, and F. C. Keil. 2006. Reflections of other minds: How primate social cognition can inform the function of mirror neurons. *Current Opinion in Neurobiology* 16:230–34.

MacIntyre, A. 1999. *Dependent Rational Animals: Why Human Beings Need the Virtues*. Chicago: Open Court.

MacLean, P. 2003. *The Triune Brain in Evolution: Role in Paleocerebral Functions*. New York: Springer.

Manger, P. 2006. An examination of cetacean brain structure with a novel hypothesis correlating thermogenesis to the evolution of a big brain. *Biological Reviews* 81:292–338.

Marino, L., R. C. Conner, R. E. Fordyce, L. M. Herman, P. R. Hof, L. Lefebvre, D. Lusseau et al. 2007. Cetaceans have complex brains for complex cognition. *PLoS Biology* 5(5). http://biology.plosjournals.org/perlserv/?request=get-document&doi=10.1371/journal.pbio.0050139&ct=1.

Markowitz, H. 1982. *Behavioral enrichment in the Zoo*. New York: Van Reinhold Company.

McComb, K., C. Moss, S. M. Durant, L. Baker, and S. Sayialel. 2001. Matriarchs as repositories of social knowledge in African elephants. *Science* 292:417–19.

McCullough, M. E. 2008. *Beyond Revenge: The Evolution of the Forgiveness Instinct*. San Francisco: Jossey-Bass.

Mech, L. D. 1970. *The Wolf.* New York: Doubleday.

Mehdiabadi, N. J., C. N. Jack, T. T. Farnham, T. G. Platt, S. E. Kalla, G. Shaulsky, D. C. Queller, J. E. Strassmann. 2006. Kin preference in a social microbe. *Nature* 442:881–82.

Melis, A., B. Hare, and M.Tomasella. 2006. Chimpanzees recruit the best collaborators. *Science* 311:1297–1300.

Mitchell, L. E. 1998. *Stacked Deck: A Story of Selfishness in America.* Philadelphia: Temple University Press.

Mitchell, R. W., and N. S. Thompson, eds. 1986. *Deception: Perspectives on Human and Nonhuman Deceit.* Albany: SUNY Press.

Moll, J., R. de Oliveira-Souza, and R. Zahn. 2008. The moral basis of moral cognition: Sentiments, concepts, and values. *Annals of the New York Academy of Sciences* 1124: 161–80.

Moll, J., R. Zahn, R. de Oliveira-Souza, F. Krueger, and J. Grafman. 2005. The neural basis of human moral cognition. *Nature Reviews: Neuroscience* 6:799–809.

Moll, J., F. Kreuger, R. Zahh, M. Pardini, R. de Oliveira-Souza, and J. Grafman, 2006. Human frontal-mesolimbic networks guide decisions about charitable donation. *Proceedings of the National Academy of Sciences* 103:15623–28.

Nichols, S. 2004. *Sentimental Rules: On the Natural Foundations of Moral Judgments.* New York: Oxford University Press.

Niteki, M. H., ed. 1990. *Evolutionary Innovations.* Chicago: University of Chicago Press.

Nowak, M. A. 2006. Five rules for the evolution of cooperation. *Science* 314:1560–63.

Nowak, M. A., and K. Sigmund. 2005. Evolution of indirect reciprocity. *Nature* 437: 1291–98.

Nussbaum, M. 2001. *Upheavals of Thought: The Intelligence of Emotions.* Cambridge: Cambridge University Press.

Packer, C. 1977. Reciprocal altruism in *Papio anubis. Nature* 265:441–43.

Panksepp, J. 1998. *Affective Neuroscience: The Foundations of Human and Animal Emotions.* New York: Oxford University Press.

———. 2003. "Laughing" rats and the evolutionary antecedents of human joy? *Physiology and Behavior* 79:533–47.

———. 2005. Beyond a joke: From animal laugher to human joy. *Science* 308:62–63.

Parr, L. A., B. M. Waller, and J. Fugate. 2005. Emotional communication in primates: Implications for neurobiology. *Current Opinion in Neurobiology* 15:1–5.

Pellis, S. 2002. Keeping in touch: Play fighting and social knowledge. In *The Cognitive Animal,* ed. M. Bekoff., C. Allen, and G. M. Burghardt, 421–27. Cambridge, MA: MIT Press.

Pfaff, D. 2007. *The Neuroscience of Fair Play: Why We (Usually) Follow the Golden Rule.* New York: Dana Press.

Poole, J. 1996. *Coming of Age with Elephants: A Memoir.* New York: Hyperion.

———. 1998. An exploration of a commonality between ourselves and elephants. *Etica & Animali* 9 (98): 85–110.

Porter, R. H., M. Wyrick, and J. Pankey. 1978. Sibling recognition in spiny mice. *Behavioral Ecology and Sociobiology* 3:61–68.

Post, S. G., L. G. Underwood, J. Schloss, and W. G. Hurlbut, eds. 2002. *Altruism and Altruistic Love: Science, Philosophy, and Religion in Dialogue.* New York: Oxford University Press.

Preston, S. D., and F. B. M. de Waal. 2002a. The communication of emotions and the possibility of empathy in animals. In *Altruism and Altruistic Love: Science, Philosophy, and Religion in Dialogue*, ed. Stephen Post et al. New York: Oxford University Press.

———. 2002b. Empathy: Its ultimate and proximate bases. *Behavioral and Brain Sciences* 25:1–72.

Raby, C. R., D. M. Alexis, A. Dickinson, and N. S. Clayton. 2007. Planning for the future by western scrub-jays. *Nature* 445:919–21.

Range, F., L. Horn, Z. Viranyi, and L. Huber. 2008. The absence of reward induces inequity aversion in dogs. *Proceedings of the National Academy of Sciences.* http://www.pnas.org/cgi/doi/10.1073/pnas.0810957105.

Reader, S. M., and K. N. Laland. 2002. Social intelligence, innovation, and enhanced brain size in primates. *Processing of the National Academy of Science* 99:4436–41.

Rice, George E., and Priscilla Gainer. 1962. "Altruism" in the albino rat. *Journal of Comparative and Physiological Psychology* 55:123–25.

Rilling, J. K., D. Gutman, T. Zeh, G. Pagnoni, G. Berns, and C. Kilts. 2002. A neural basis for social cooperation. *Neuron* 25:395–405.

Rizzolatti, G., and L. Craighero. 2004. The mirror-neuron system. *Annual Review of Neuroscience* 27:169–92.

Ross, Marina D., Susanne Menzler, and Elke Zimmermann. 2008. Rapid facial mimicry in orangutan play. *Biology Letters* 4:27–30. http://journals.royalsociety .org/content/?k=davila+ross.

Roth, G., and U. Dicke. 2005. Evolution of the brain and intelligence. *Trends in Cognitive Science* 9:250–57.

Rottschaefer, W. A. 1998. *The Biology and Psychology of Moral Agency.* Cambridge: Cambridge University Press.

Rutte, C., and M. Taborsky. 2007. Generalized reciprocity in rats. *PLoS Biology* 5 (7): e196.

Sanfey, A. G., J. Rilling, J. Aronson, L. Nystrom, and J. Cohen. 2003. The neural basis of economic decision-making in the ultimatum game. *Science* 300:1955–58.

Sapolsky, R. 2004. *Why Zebras Don't Get Ulcers.* 3rd ed. New York: Holt Paperback.

Sapolsky, R. M. 2002. *A Primate's Memoir.* New York: Touchstone Books.

Schaller, G. B., and G. R. Lowther. 1969. The relevance of carnivore behavior to the study of early hominids. *Southwestern Journal of Anthropology* 25:307–41.

Schuster, R. 2002. Cooperative coordination as a social behavior: Experiments with an animal model. *Human Nature* 13:47–83.

Seed, A., N. Clayton, and N. Emery. 2008. Cooperative problem solving in rooks (*Corvus frugilegus*). *Proceedings of the Royal Society B*, DOI: 10.1098/rspb.2008.0111.

Serpell, J. 1996. *In the Company of Animals: A Study of Human-Animal Relationships.* Cambridge: Cambridge University Press.

Seymour, B., T. Singer, and R. Dolan. 2007. The neurobiology of punishment. *Nature Reviews: Neuroscience* 8:300–309.

Shapiro, P. 2006. Moral agency in other animals. *Theoretical Medicine* 27:357–73.

Sherman, P. 1977. Nepotism and the evolution of alarm calls. *Science* 197:1246–53.

Shermer, M. 2004. *The Science of Good and Evil*. New York: Henry Holt and Company.

Silk, J., S. F. Brosnan, J. Vonk, J. Henrich, D. J. Povinelli, A. S. Richardson, S. P. Lambeth, J. Mascaro, and S. J. Schapiro. 2005. Chimpanzees are indifferent to the welfare of unrelated group members. *Nature* 437:1357–59.

Silk, J. B., R. M. Seyfarth, and D. L. Cheney. 1999. The structure of social relationships among female savanna baboons in Moremi Reserve, Botswana. *Behaviour* 136: 679–703.

Simmonds, M. P. 2006. Into the brains of whales. *Applied Animal Behaviour Science* 100: 103–16.

Singer, T., B. Seymour, J. P. O'Doherty, K. E. Stephen, R. J. Dolan, and C. D. Frith. 2006. Empathic neural responses are modulated by the perceived fairness of others. *Nature* 439:466–69.

Siviy, S. 1998. Neurobiological substrates of play behavior: Glimpses into the structure and function of mammalian playfulness. In *Animal Play: Evolutionary, Comparative, and Ecological Perspectives*, ed. M. Bekoff and J. A. Byers, 221–42. New York: Cambridge University Press.

Slater, K. Y., C. M. Schaffner, and F. Aureli. 2007. Embraces for infant handling in spider monkeys: Evidence for a biological market? *Animal Behaviour* 74:455–61.

Smith, John Maynard. 1982. *Evolution and the Theory of Games*. Cambridge: Cambridge University Press.

Sober, E., and D. S. Wilson. 1998. *Unto Others: The Evolution and Psychology of Unselfish Behavior*. Cambridge, MA: Harvard University Press.

Solomon, R. C. 1995. *A Passion for Justice*. Lanham, MD: Rowman & Littlefield.

Sorabji, R. 1993. *Animal Minds and Human Morals: The Origins of the Western Debate*. Ithaca: Cornell University Press.

Spinka, M., R. C. Newberry, and M. Bekoff. 2001. Mammalian play: Training for the unexpected. *Quarterly Review of Biology* 76:141–68.

Steiner, G. 2005. *Anthropocentrism and Its Discontents: The Moral Status of Animals in the History of Western Philosophy*. Pittsburgh: University of Pittsburgh Press.

Stevens, Jeffrey R., and Marc D. Hauser. 2004. Why be nice? Psychological constraints on the evolution of cooperation. *Trends in Cognitive Sciences* 8:60–65.

Subiaul, Francys, Jennifer Vonk, Sanae Okamoto-Barth, and Jochem Barth. 2008. Do chimpanzees learn reputation by observation? Evidence from direct and indirect experience with generous and selfish strangers. *Animal Cognition* DOI 10.1007/s10071=008-0151-6.

Sussman, R. W., P. A. Garber, and J. M. Cheverud. 2005. Importance of cooperation and affiliation in the evolution of primate sociality. *American Journal of Physical Anthropology* 128:84–97.

Talmi, D., and C. Frith. 2007. Feeling right about doing right. *Nature* 446:865–66.

Tancredi, L. 2005. *Hardwired Behavior: What Neuroscience Reveals about Morality.* Cambridge: Cambridge University Press.

Taylor, S. 2002. *The Tending Instinct: How Nurturing Is Essential for Who We Are and How We Live.* New York: Henry Holt and Company.

Thayer, B. A. 2004. *Darwin and International Relations: On the Evolutionary Origins of War and Ethnic Conflict.* Lexington: University of Kentucky Press.

Tinbergen, N. 1951/1989. *The Study of Instinct.* New York: Oxford University Press.

———. 1963. On aims and methods of ethology. *Zeitschrift für Tierpsychologie* 20: 410–33.

Trivers, R. L. 1971. The evolution of reciprocal altruism. *Quarterly Review of Biology* 46:35–57.

Turiel, E., M. Killen, and C. Helwig. 1987. Morality: Its structure, functions, and vagaries. In Kagan and Lamb 1987, 155–244. Chicago: University of Chicago Press.

Warneken, F., B. Hare, A. P. Melis, D. Hanus, and M. Tomasello. 2007. Spontaneous altruism by chimpanzees and young children. *PLoS Biology* 5(7): e184.

Watson, D. M., and D. B. Croft. 1996. Age-related differences in playfighting strategies of captive male red-necked wallabies (*Macropus rufogriseus banksianus*). *Ethology* 102:336–46.

Wechkin, S., J. H. Masserman, and W. Terris, Jr. 1964. Shock to a conspecific as an aversive stimulus. *Psychonomic Science* 1:17–18.

Wegner, D. M. 2002. *The Illusion of Conscious Will.* Cambridge, MA: MIT Press.

Wemelsfelder, F., and A. B. Lawrence. 2001. Qualitative assessment of animal behaviour as an on-farm welfare-monitoring tool. *Acta Agriculturae Scandinavica* 30: S21–S25.

Wemmer, C., and C. A. Christen, eds. 2008. Elephants and ethics: Toward a morality of coexistence. Baltimore: The Johns Hopkins University Press.

West, Stuart A., Ido Pen, and Ashleigh S. Griffin. 2002. Cooperation and competition between relatives. *Science* 296:72–75.

White, T. I. 2007. *In Defense of Dolphins: The New Moral Frontier.* Malden, MA: Blackwell Publishing.

Wilkinson, G. 1984. Reciprocal food sharing in vampire bats. *Nature* 308:181–84.

———. 1987. Reciprocal altruism in bats and other mammals. *Ethology and Sociobiology* 9:85–100.

Wilkinson, R. 2007. *Unhealthy Societies: The Affliction of Inequality.* Oxford: Taylor & Francis.

Wilson, E. O. 1975. *Sociobiology: The New Synthesis.* Cambridge, MA: Belknap.

———. 1978. *On Human Nature.* Cambridge, MA: Harvard University Press.

Wilson, J. Q. 1993. *The Moral Sense.* New York: The Free Press.

Wilson, T. 2002. *Strangers to Ourselves: Discovering the Adaptive Unconscious.* Cambridge, MA: Belknap/Harvard University Press.

Zahn-Waxler, C., M. Radke-Yarrow, E. Wagner, and M. Chapman. 1992. Development of concern for others. *Developmental Psychology* 28:126–36.

INDEX

Babyl (elephant), 102

Balcombe, Jonathan, 45, 96–97, 125, 126

Baldwin, Ann, 97

Barnard, Neal, 96

Barrett, Louise, 73

Bates, Lucy, 37

bats: altruism, 7; midwifery, 136–37, 145, 152; reciprocity, 73

Batson, Daniel, 134

bears, greater relative brain and neocortex size than social carnivores, 51

bees, prosocial behavior, 13, 47

behavior, ultimate and proximate explanations for, 26

behavioral flexibility, 79, 83; and intelligence, 48; and theories of social evolution, 36

"Behavioural Reactions of Elephants towards a Dying and Deceased Matriarch" (Douglas-Hamilton et al.), 103–4

Bekoff, Marc, xiii; on animal cooperation, 58; on anthropomorphism, 40, 43; coyote studies, 130; The Emotional Lives of Animals, 40, 45; on fairness, 115; observations of dog empathy, 108; observations of dogs playing, 117, 119; observations of elephant empathy, 102; on patterns of antipredatory scanning, 91; research on play bows, 122, 123; on stress experienced by laboratory animals, 97; view of animal play, 118

Belding's ground squirrels, 68, 69

"bias to cooperate," 76

Biggs, Lorraine, 60

Binti Jua (female western lowland gorilla), 1–2, 37, 108

biological determinism, and morality, 21–23

biology, 151; conservatism in labeling behavior, 38; and ethics, mixing of, 137; technical meaning of spite in,

61. See also evolutionary biology; neurobiology; sociobiology

body mimicry, 89

Boehm, Christopher, 146–47

bonobos, 108; cooperative behavior, 81–82; spindle cells, 101; and threshold requirements for morality, 83

bottlenose dolphins, 52, 94

Boysen, Sarah, 111

Bradshaw, Gay, 105, 106

brain size: brain-size-to-sociality correlations, 51; of cetaceans, 52; and social group size, 50–51

Brosnan, Sarah, 6, 111, 127

Bshary, Redouan, 70

Burghardt, Gordon, 117

by-product mutualism, 69

Byrne, Richard, 37

canids: conflict resolution, 81; importance of play for development of social skills and social bonds, 117; play behaviors from different contexts, 119; play bow, 121, 122, 123

captive animals, study of, 27

capuchin monkeys: food sharing among, 56; inequity aversion, 6, 111, 127–28

caring, 8, 87

Cartesian view of animals, 10–11

cat, domestic, 53

catecholamine hormones, 45

caudate nucleus, and trust, 39

causation, of behavior, 26

cetaceans: brain size of, 52; empathy in, 90, 94, 101–2; evidence for moral behavior in, 9; grieving in, 102. See also dolphins; whales

c-FOS, 119

chacma baboons, 73

chaffinches, cooperation among, 58

Chapman, Audrey, 47, 130

Cheney, Dorothy, 71

Cheverud, James, 4, 57, 70

chimpanzees, xii; capacity for character judgments, 114–15; cooperative behavior, 64, 81–82; empathy, 6, 97–98; food sharing among, 56; generalized reciprocity, 75; inequity aversion, 111, 128–29; reciprocity and fairness, 19; resource-sharing, 111–12; self-protection groups among females, 58; spindle cells, 101; and threshold requirements for morality, 83; and "ultimatum game," 110–11

Chimpanzees of Gombe, The (Goodall), 4

chordates, 88

Church, Russell, 96, 145

civil rights movement, 148–49

Clayton, Nicola, 55

clusters, behavioral, 8

Clutton-Brock, Timothy, 75

cognitive complexity, 83, 145

cognitive ethology, xiv, 24, 25–27, 44

cognitive neuroscience, 32

cognitive psychology, 151

cognitive speciesism, 80–82

colonial microorganisms, 46

communal care of young, 58, 59

communal nursing, 58

community concern, 48

companion animals, relationship with human guardians, 108

comparative research between and among species, 81

compassion, xiv, 5, 8, 87

competition among animals, balancing with cooperation, 57–58, 79

competition paradigm, hegemony of, xii

conflict negotiation, 59

conscience: in animals, 145–47; and Charles Darwin, 145–46; in humans, as response to shift from subsistence to hunting societies, 146–47; as internalized norms, 146; and morality, 138, 143

consciousness, representative of range of processes and behaviors, 39

consolation, xiv, 8, 87, 100

conventional violations, 15

Cools, Annemieke, 81

cooperation, 55–57; allows specialization and promotes biological diversity, 59; and anger, 78; balancing with competition, 57–58; binds and maintains social ties, 56; and capacity to compare, 133; cognitive foundations of, 78–80; evolution of, 66–76; interdisciplinary approach to, 75; joint defense of territory or resources, 70; many different reasons for, 58–59; and moral behavior, 57; as moral behavior, 82–84; neurological mechanisms underlying, 63, 107; no special meaning in biology, 61–62; not always reciprocal, 74; often paired with affiliative behavior, 63; problem in labeling, 62; psychological mechanisms underlying, 76–77; some elements of necessary for justice, 133; sometimes equated with altruism, 62; tied to larger group of prosocial affiliative and helping behaviors, 62; ultimate and proximate explanations for, 65–66; among unrelated individuals, 69; various levels of, 63–64. *See also* mutualism

cooperation cluster, xiv, 8, 54, 59–60, 138, 151

cooperation skeptics, 64

cooperative hunting, 59, 63, 64, 65, 70

corals, 46

core behavior patterns, 149, 151

cortisol, 48, 93

corvids, 10, 81

Costa, James, 46

coyotes: cooperation of ravens with, 56; cubs playing, 118; importance of play for development of social skills and

impairment, 106; effect of loss of matriarch on society, 106; empathy in, 6–7, 84, 90, 94, 102–5; evidence for moral behavior in, 9; family groups, xv; grieving for dead, 104; interest in corpses and bones, 104–5; killing of rhinoceroses by cull orphans, 106; post traumatic stress disorder (PTSD), 106; social breakdown in societies of, 105–6; and threshold requirements for morality, 83

Emery, Nathan, 55

emotional complexity, 145

emotional contagion, 21, 89, 90–92

emotional empathy, 89. *See also* empathy

Emotional Life of Animals, The (Bekoff), 40, 45

"Emotional Reactions of Rats to the Pain of Others" (Church), 96

emotions, in animals: and fellow feelings, 44–45; recognizing, 43–44; research into, x

empathic ethology, 34–36

empathy: adaptive, 90–92; in adult non-primate mammals, 86; and altruism, 92, 109, 134; in apes, 99–100; basic prosocial behavior, 107; in bottle-nose dolphins, 94; as building block of morality, 107; in cetaceans, 90, 94, 101–2; in chimpanzees, 6, 97–98; as class of behavior patterns across species, 89; cognitive, 21, 89, 90, 100; and cooperation, 78, 134; costs of, 92–93; and cruelty, 17; defined, 87–88; in diana monkeys, 6; in dolphins, 102, 108; in elephants, 6–7, 84, 90, 94, 102–5; as emotional linkage, 88, 89; evidence of in animals through anecdotes and scientific study, 5, 85–87; facial ecology of, 93–94; as intersubjective experience, 88; and

justice, 109, 134; in mice, 21, 85, 86, 94, 96–97; and mirror neurons, 100–101; in monkeys, 98–100; and mother-infant bond, 107; neuroscience of, 32, 95, 100–101, 134; notion of as nested levels, 89; in primates, 94, 97–100; in rats, 94; in rhesus monkeys, 98–99; in rodents, 94, 96–97; simple to complex, 88–90; in social carnivores, 94; across species, 107–9; in toothed whales, 94; use of term in animal behavior field, 11

"empathy-altruism" hypothesis, 134

empathy cluster, xiv, 8, 54, 87–88, 138, 151

endogenous opioid peptides (EOPs), 63, 77

envy, and cooperation, 78

episodic memory, 51

equity, xiv, 8

Erhlich, Paul, xiii

ethics, and biology, mixing of, 137

ethology, 25, 54, 151; conservatism in labeling behavior, 38; evidence for animal cooperation, 64; evidence for animal empathy, 94; fieldwork, 97; predominance of competition paradigm in, xii; reliance on data drawn from observation, 28; search for similarities and differences across species, 38–39; view of play, 118

eusocial arrangements, 47

evolution: of a behavior, 26; and cooperation, 59

evolutionarily stable strategy (ESS), 115

evolutionary biology: concept of empathy, 88; predominance of competition paradigm in, xii

evolutionary continuity, xi, 39, 41, 94, 97, 137, 139, 141

evolutionary game theory, 75

evolutionary psychology, 22

habits of mind, 10–11
Hamilton, William D., 67, 69, 78
Hamlin, Kiley, 114
Hare, Brian, 75, 81
Harlow, Harry, 99
harm and benefits, as moral currency, 14
Hart, Ben, 73
Hart, Lynette, 73
Hauser, Marc, 79, 80
Heinrich, Bernd, 10, 56, 81
helping, 8, 87
Highfield, Roger, 92
Hinde, Robert, 146
Hof, Patrick, 29–30
Hoffman, Martin, 150
Holekamp, Kay, 51–52
hominids: brain evolution, 140; resemblance of social carnivore behavior to early, 9
hominoids: cognitive empathy in, 90; spindle cells in, 101
honesty, xiv, 17, 59
Hornaday, William, ix, 15
Horowitz, Alexandra, 42, 120
huddling, 70
Hudson, Richard, 82
human immunodeficiency virus (HIV), 67
humans: babies, social intelligence, 114; concept of animal morality as threat to, 141–42; ethical responsibilities toward animals, 149–50; large prefrontal cortex, 147; laughter, 92; multifaceted intelligence, 50; responses to inequity, 130; self-consciousness about grounds of actions, 140. See also morality, human
human-to-animal inference, 40
humpback whales, spindle cells, 101
Humphrey, Nicholas, 50
hunting, cooperative, 59, 63, 64, 65, 70
hyenas: evidence for moral behavior in, 9; spotted, cooperation among, 55–

56; and threshold requirements for morality, 83

immersion research, 35
immorality, and morality, 15–19
impala, allogrooming among, 73
"impartial spectator" role, 132, 133
impulse control, and moral behavior, 146
indignation, 8, 113, 115
individual selection, xiv
industrial farming, 150
inequity, human responses to, 130
inequity aversion, in animals, 6, 111, 115, 127–29
infanticide, 17
insect sociality, 46–47
instinct, 144
intelligence: and behavioral flexibility, 48; defined, 48, 49; and morality, 48–50; multiple, 50; orthogenetic view of, 49; representative of range of processes and behaviors, 39; research into in animals, x; species-specific, 49; view of independent evolution in different classes of vertebrates, 49
intentional control, 140
invasive laboratory research, 97
invertebrates, 46

Jackson, Philip, 93
Jacob and Esau, 68
Jensen, Keith, 110, 111, 112
Johnson, Dominic, 72
joint defense of territory or resources, 70
Jolly, Allison, 50
Joni (chimpanzee), 109
joy, and morality, 126
justice: as continuous and evolved trait, 115; defined, 113; evolution out of cooperation and altruism, 134; generally viewed as abstract principle,

for moral behavior in, 9; food sharing among, 56; gratitude and indignation, 128; inequity aversion, 6, 111, 127–28; mirror neurons in, 28–29, 95, 100; reciprocity in, 73, 128; wire-monkey experiments, 99

Montague, Reed, 39

moral agency: and animals, 143; philosophical definition, 144; species-specific and context-specific, 144

"Moral Agency in Other Animals" (Shapiro), 132

moral agent, vs. moral patient, 144

moral complexity: and social complexity, xiii; variation among species, 20

morality: as affective foundation of cooperation, 76–77; behavioral clusters, xiv; and biological determinism, 21–23; descriptive accounts of, 148; difficulty of defining, 11; evolution of, xi, xiii, 61, 149; and evolution of sociality, xiii, 45; expansive definition of, 32; extension of term to include nonhuman animals, 137–38; and immorality, 15–19; and intelligence, 48–50; and manners, 14–15; meaning of, 138; nested levels of increasing complexity, 139; normative accounts of, 148; not primarily abstract, 132; other-regarding, 138, 148; patterns of behavior unique to humans, 139–40; and prosociality, 12–13, 14, 138, 148; questions about defining characteristics of, 143; as social phenomenon, 7; species-relative view of, xii, xiv, 19–21, 39, 144, 147–49; as system of behavioral control, 15; Western philosophical accounts of, 149; working definition of, 7. *See also* morality, animal; morality, human

morality, animal: challenge to stereotypes about animals, 11; and codes of conduct, 5; and complex cooperative behavior, 83; defined, 54; different in degree from that of humans, 139; disciplinary foundations for, 24; displays of, ix–x; foundations of, 45–47; and moral agency, 143–45; philosophical questions about, 137; possible objections to, 142–43; scientific resistance to term, 31; as strategy for social living, x, 3, 45; as threat to concept of human uniqueness, 11, 141–42; threshold requirements for, 13, 83, 145; ties to sociality, intelligence, and emotion, 25; and unified research agenda, 54, 152

morality, human: conditioned and instinctive, 145; convergence with animal morality, 11, 31; and law and religion, 15; neural basis of, 151; uniqueness, 132

moral philosophy, 24, 32–33, 137

moral psychology, xiv, 15, 24

moral relativism, 147

moral violations, 15

moray eels, 70

mother-infant bond, and development of empathy, 107

mountain sheep, play fighting of cubs, 117

mounting behavior, 117

Muller, Martine, 17

multifactorial hypothesis, 51

multiple intelligences, 50

Mundebvu (elephant), 105

mutation, 59

Mutual Aid (Kropotkin), 57

mutualism, 65, 67, 69–70, 71

narrative, defined, 37

narrative ethology, 36–38

natural selection, 3, 27, 59, 140

Nelissen, Mark, 81

Nell, Victor, 18

neocortex: functions of, 89; volume of, and group size, 50–51

neo-Darwinian thought, 21–22

neurobiology, 151

neuroscience, 54; evidence for animal empathy, 94; and foundations of empathy and fairness, 95, 134; view of play, 118

Newberry, Ruth, 118

New World monkeys, affiliative social interactions, 4

norm, defined, 14

normative accounts of morality, 148

normative self-government, 140

Norway, egalitarianism and health in, 130

Nowak, Martin, 59

numerical quantification, 80

observed behavior, translation into scientific language, 38

Old World monkeys, affiliative social interactions, 4

olfactory recognition, 68–69

On Aggression (Lorenz), 25–26

On Human Nature (Wilson), 21

ontogeny, of a behavior, 26

orangutans: emotional contagion in, 92; spindle cells, 101

orcas, grieving in, 102

Origins of Sociality, The (Chapman and Sussman), 47

orthogenetic view of intelligence, 49

Other Insect Societies, The (Costa), 46–47

other minds, problem of, 44

other-regarding behaviors, 13, 14, 138, 148

oxytocin, 32, 63, 77–78

pair-bonding, 77

paleomammalian brain, 89

Pankey, Jan, 68

Panksepp, Jaak, 28, 77, 86, 87, 126

Pan troglodytes, 112

parasitic wasp, 61

Passion for Justice, A (Solomon), 112, 132

Pellis, Sergio, 119, 124

Pen, Ido, 68

Pfaff, Donald, 28

philosophy, 54, 151; and animal morality, xiv, 30–33, 137, 149; concept of empathy, 88; definition of moral agency in, 144; resistance to use of term "moral" in relation to nonhuman animals, 10

Pierce, Jessica, xiii

Pitcher, Tony, 73

plasticity, 145

play. *See* social play behavior

play bow, 121, 122, 123, 126–27

play fighting, 117

play pant/laugh, 92, 120

Pleasurable Kingdom (Balcombe), 45

pleasure: and justice, 113, 115; and morality, 126; and play, 125–26

polar bears, affection display, 46

Poole, Joyce, 6–7, 102

Porter, Richard, 68

post traumatic stress disorder (PTSD), in elephants, 106

prairie dogs, 91

"predator inspection," in fish, 73–74

predatory behavior, 122

prefrontal cortex: and conscience, 147; human, 139–40; spindle cells, 101

Preston, Stephanie, 88, 89, 90, 101, 138, 139

primate paradigm, 81

primates, nonhuman: affiliative rather than agonistic social behavior, 57–58; empathy in, 94, 97–100; inequity aversion, 127–29; moral behavior in, 9; social attachment in, 99; social interactions mostly affiliative rather than agonistic, 4; spindle cells, 101. *See also* apes, empathy in; great apes; monkeys

Primate's Memoir, A (Sapolsky), 16

Schore, Allan, 106
Schuster, Richard, 76
science, ambivalence in use of term "moral" in relation to nonhuman animals, 10, 11
scrub jays, episodic memory, 51
Seed, Amanda, 55
selection pressures, 52
self-assessment, 147
self-control, 147
self-handicapping ("play inhibition"), 119, 123–24, 125
self-interest, moving beyond, and trust, 131
Selfish Gene, The (Dawkins), 61
"selfish genes," theories of, 61
selfishness: ambiguity of in relation to animal morality, 60–61; use of term in animal behavior field, 11
self-regarding behaviors, 13–14
"sense of justice," 132
sentinel behavior, 13, 14, 69, 70
Seyfarth, Robert, 71
shame, 147
Shapiro, Paul, 144
"shared-state mechanism," 88
sharing, xiv, 8, 56, 111–12
Sherman, Paul, 68
short-term studies, 35
siblicide, 3, 5
Siegel, Sarita, 41
Silk, Joan, 71
Simmonds, Mark, 101–2
Singer, Tania, 134
situational morality, 148
Siviy, Stephen, 118–19
Slater, Kathy, 73
slime molds, 46, 82–83
Sober, Elliott, 60
social attraction factor, 4
social carnivores: cognitive empathy, 90, 94; evidence for moral behavior in, 9
social cognition, 79

"Social Communication in Canids" (Bekoff), 122
social competition factor, 5
social complexity, and moral complexity, xiii, 53
social evaluation, as foundation for moral behavior, 114
social groups: and compromises in individual freedom, 130–31; small tightly knit, 63; and social hierarchies, 47; stability in, and social play behavior, 119–20; system of proscriptions and prohibitions, 13; trust in, 17
social homeostasis, xi, 126
social insects, 46
Social Intelligence (Goleman), 47
social intelligence, 47–48
social intelligence hypothesis, 50–53; limitations and counter-examples, 51; and moral behavior, 52–53
sociality, evolution of, xii, 36, 45–47, 115. See also social groups
social mammals, 46, 47
social neuroscience, xiv, 24, 27–30, 44
social play behavior, xii, 59, 116–20; adaptive, 117–18; and apology and forgiveness, 126–27; and behavioral flexibility, 118; and behaviors from other contexts, 122; as "brain food," 118; contagion of, 91–92; egalitarianism, 121, 123–25; evolutionary roots of, 117; and fairness and trust, 112, 116, 120–21, 124–25; and group stability, 119–20; and honesty, 123; and joy and pleasure, 120; and logical reasoning, 118; and morality, 115–16, 121–23; norms, 14; play bow, 121, 122, 123, 126–27; play fighting, 117; play-invitation signals, 119, 124, 125, 134; play pant/laugh, 92, 120; and pleasure, 125–26; and reproductive fitness, 129–30; in social carnivores, 5–6; as training for the unexpected, 118